GCSE Success

WORKBOOK

Physics

**Colin Porter
and Charles Cotton**

Contents

Revised

Energy
- 4 The Electricity Supply
- 6 Generating Electricity
- 8 Renewable Sources of Energy
- 10 Electrical Energy and Power
- 12 Electricity Matters
- 14 Particles and Heat Transfer

Waves
- 16 Describing Waves
- 18 Wave Behaviour
- 20 Seismic Waves and the Earth

Electromagnetic Waves
- 22 The Electromagnetic Spectrum
- 24 Light, Radio Waves and Microwaves
- 26 Wireless Communications 1
- 28 Wireless Communications 2
- 30 Infrared
- 32 The Ionising Radiations
- 34 The Atmosphere

Beyond the Earth
- 36 The Solar System
- 38 Space Exploration
- 40 A Sense of Scale
- 42 Stars
- 44 Galaxies and Red-Shift
- 46 Expanding Universe and Big Bang

Forces and Motion
- 48 Distance, Speed and Velocity
- 50 Speed, Velocity and Acceleration
- 52 Forces
- 54 Acceleration and Momentum
- 56 Pairs of Forces: Action and Reaction
- 58 Work and Energy
- 60 Energy and Power

Electricity

- 62 Electrostatic Effects
- 64 Uses of Electrostatics
- 66 Electric Circuits
- 68 Voltage or Potential Difference
- 70 Resistance and Resistors
- 72 Special Resistors
- 74 The Mains Supply

Radioactivity

- 76 Atomic Structure
- 78 Radioactive Decay
- 80 Living with Radioactivity
- 82 Uses of Radioactive Materials
- 84 Nuclear Fission and Fusion

Light

- 86 Refraction, Dispersion and TIR
- 88 Lenses
- 90 Seeing Images
- 92 Telescopes and Astronomy

Further Physics

- 94 Electromagnetic Effects
- 96 Kinetic Theory
- 98 The Gas Laws
- 100 Medical Physics

The Electricity Supply

Multiple-choice questions

Choose just one answer: A, B, C or D.

1. Which of the following is a primary energy resource? (1 mark)
   - A mains electricity
   - B uranium fuel in a nuclear power station
   - C heat from a radiator
   - D the kinetic energy of a turbine

2. What type of diagram can be used to illustrate energy transfers? (1 mark)
   - A Gantt chart
   - B ray diagram
   - C Sankey diagram
   - D Venn diagram

3. Which of the following can be used to generate electricity without the need for movement? (1 mark)
   - A tidal energy
   - B geothermal energy
   - C chemical energy in natural gas
   - D solar energy

4. Which of the following is a secondary energy resource? (1 mark)
   - A electrical energy from a battery
   - B nuclear energy
   - C coal
   - D wave energy

5. In a petrol-driven engine there are many energy transfers taking place. Which type of energy is considered to be wasted energy? (1 mark)
   - A electrical
   - B nuclear
   - C heat
   - D kinetic

Score / 5

Short-answer questions

1. a) Draw a Sankey diagram to show the energy transfers that take place in a light bulb with an input energy of 100 J and an efficiency of 10%. (3 marks)

   b) For the diagram above, state the type of energy for each of the following: (3 marks)

   i) input ....

   ii) useful output ....

   iii) wasted ....

2. For every 1 kg of coal burned in a power station, 8.4 MJ of electrical energy and 15.6 MJ of heat energy are produced. Calculate the % efficiency of electricity generation. (2 marks)

Score / 8

GCSE-style questions

Answer all parts of all questions. Continue on a separate sheet of paper if necessary.

1 With traditional fuels, such as coal, being in finite supply there is much interest in developing renewable energy resources. One possible source is solar energy.

a) There are two ways of using solar energy - by installing 'passive' or 'active' solar panels.

Describe the difference between the 'passive' and 'active' harnessing of solar energy. **(2 marks)**

b) State one advantage and one disadvantage of using solar energy in the UK. **(2 marks)**

c) Commercial solar panels are made from lots of 'cells'. On average, each cell has an efficiency of 15%. State what is meant by the term 'efficiency'? **(1 mark)**

d) Globally, the average electrical output of a coal-burning power station is 310 J for every 1 kJ of energy input as fuel.

Calculate the average efficiency of a coal-burning power station. **(1 mark)**

e) i) 'Energy is always conserved.' Describe what this statement means. **(1 mark)**

ii) If energy is always conserved, describe why the energy output by power-stations and solar cells as electricity does not match the total energy input. **(2 marks)**

f) The EU has set a target of increasing the use of renewable energy to 20% by 2020. Suggest two issues that need to be addressed to help achieve this target. **(2 marks)**

Score / 11

How well did you do?

0–6 Try again | 7–12 Getting there | 13–18 Good work | 19–24 Excellent!

For more information on this topic, see pages 4–5 of your Success Revision Guide.

Generating Electricity

Multiple-choice questions

Choose just one answer: A, B, C or D.

1. Which of the following energy resources is a fossil fuel? **(1 mark)**
   A oil
   B nuclear
   C wind
   D geothermal

2. Which of the following sequences correctly describes the energy transfers taking place in a coal-fired power station? **(1 mark)**
   A heat – chemical – kinetic – electrical
   B chemical – kinetic – heat – electrical
   C heat – chemical – kinetic – electrical
   D chemical – heat – kinetic – electrical

3. Which of these devices produces electricity? **(1 mark)**
   A bulb
   B transformer
   C dynamo
   D resistor

4. Which of the following non-renewable energy resources does not emit carbon dioxide gas into our atmosphere during energy transfer? **(1 mark)**
   A oil
   B nuclear
   C coal
   D natural gas

5. Which of the following is not needed for electromagnetic induction to take place? **(1 mark)**
   A coil
   B movement
   C magnet
   D heat

Score / 5

Short-answer questions

1. a) This diagram represents a coal-burning power station. Add labels to complete the diagram. **(4 marks)**

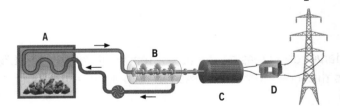

   b) Describe the functions of the following devices on the diagram above: **(3 marks)**

   i) Device B _____

   ii) Device C _____

   c) Describe why coal, oil and gas are called fossil fuels. **(1 mark)**

   _____

Score / 8

GCSE-style questions

Answer all parts of all questions. Continue on a separate sheet of paper if necessary.

1 The Government is considering opening a new power station in the highlands of Scotland.

a) One of the main objectives is to ensure that the power station does not emit carbon dioxide. A local councillor suggests using biofuels.

　i) Describe why biofuels are **not** a suitable option. (1 mark)

　ii) Suggest why the councillor might have thought that biofuels would be suitable. Explain your answer. (3 marks)

　iii) State two types of biofuels. (2 marks)

b) List three important factors that must be considered when choosing a suitable site for a new power station. (3 marks)

c) Ring the correct options to complete this explanation of the role of generators in the new power station.

The generator will induce a **voltage / current / dynamo** using coils of wire and an electromagnet. The electromagnet **remains stationary / rotates / moves up and down** inside the coils of wire so that the coils are in a **positive / negative / changing** magnetic field. (3 marks)

d) The Government decides that nuclear fuel is the most suitable fuel for the power station. They know that this will raise some concerns from the local population and want to prepare for this before announcing their proposal. Suggest two specific factors they should research before making their announcement. (2 marks)

Score　/ 14

How well did you do?

0–8 Try again　　9–14 Getting there　　15–20 Good work　　21–27 Excellent!

For more information on this topic, see pages 6–7 of your Success Revision Guide.

Renewable Sources of Energy

Multiple-choice questions

Choose just one answer: A, B, C or D.

1 Which of the following is NOT a renewable energy resource? **(1 mark)**
   A geothermal energy
   B nuclear fuel
   C biofuel
   D solar energy

2 Which of these renewable energy resources emits carbon dioxide gas into the atmosphere during energy transfer? **(1 mark)**
   A biofuels
   B geothermal energy
   C tides
   D wind

3 Which of the following renewable energy resources is not dependent on the weather? **(1 mark)**
   A solar energy
   B hydroelectric power
   C tidal energy
   D wind power

4 Which of the following renewable energy resources does not make use of the Sun's energy? **(1 mark)**
   A solar energy
   B wave power
   C geothermal energy
   D wind power

5 Which of the following renewable energy resources is influenced by the Moon? **(1 mark)**
   A tidal energy
   B solar energy
   C geothermal energy
   D wind power

Score / 5

Short-answer questions

1 Explain what is meant by 'renewable' and 'non-renewable' in the context of energy resources. **(2 marks)**

2 List three renewable energy resources and three non-renewable energy resources. **(6 marks)**

3 For each of the following types of renewable energy, state one possible disadvantage. **(5 marks)**

   a) hydroelectric power

   b) tidal power

   c) solar energy

   d) wind power

   e) geothermal energy

Score / 13

GCSE-style questions

Answer all parts of all questions. Continue on a separate sheet of paper if necessary.

1 Hydroelectric power stations can be used to store surplus energy during periods of low demand by pumping water back into the reservoir.

  a) Explain how, during subsequent periods of high demand, hydroelectric power stations respond rapidly to meet energy requirements. **(3 marks)**

  b) Describe the energy transfers that take place in a hydroelectric power station. **(2 marks)**

2 It is proposed to build a tidal barrage on a river estuary. The local population has mixed reactions to the proposal.

  a) Suggest two reasons why some people may object to the proposal. **(2 marks)**

  b) Suggest two reasons why some people may be in favour of the proposal. **(2 marks)**

3 Explain why the Sun can be considered to be the original energy source of most of our energy resources. Make reference to specific renewable and non-renewable resources in your answer. **(6 marks)**

Score / 15

How well did you do?

| 0–10 | Try again | 11–18 | Getting there | 19–25 | Good work | 26–33 | Excellent! |

For more information on this topic, see pages 8–9 of your Success Revision Guide.

Electrical Energy and Power

Multiple-choice questions

Choose just one answer: A, B, C or D.

1 Which of these formulae shows the relationship between energy and power? **(1 mark)**
- A power = current × voltage
- B energy = power × time
- C voltage = power + energy
- D current = (power × time) / energy

2 How much energy is converted at a power of 10 W in 10 s? **(1 mark)**
- A 100 J
- B 50 J
- C 200 J
- D 500 J

3 What is the power of a device that does 1000 J of work in 10 s? **(1 mark)**
- A 1000 W
- B 50 W
- C 100 W
- D 200 W

4 How long in seconds would it take to convert 80 000 J of energy at a power of 4 kW? **(1 mark)**
- A 100 s
- B 20 s
- C 200 s
- D 10 s

5 How many kWh of energy are converted by a device operating at 100 000 W for 30 minutes? **(1 mark)**
- A 50 kWh
- B 3333 kWh
- C 0.0003 kWh
- D 30 kWh

Score / 5

Short-answer questions

1 How many joules equal 1 kWh? Give a calculation to support your answer. **(2 marks)**

2 Calculate how many joules of electrical energy are converted by a 100 W bulb in 5 minutes. **(2 marks)**

3 State the standard unit used to measure each of the following, including the symbol. **(3 marks)**

   a) current _____ b) voltage _____ c) power _____

4 State three energy saving measures that your family could take to reduce the household electricity bills. **(3 marks)**

Score / 10

GCSE-style questions

Answer all parts of all questions. Continue on a separate sheet of paper if necessary.

1 The European Union (EU) has banned the sale of 100 W filament bulbs because about 90 per cent of the input energy is wasted as heat. Instead, they are encouraging the use of more energy efficient bulb designs.

a) If electricity costs 10p a unit (kWh), how much would it cost to keep a 100 W light bulb on for 12 hours each day for a week? (2 marks)

b) The recommended replacement for a 100 W filament bulb is an 'energy saving' bulb with an efficiency of 56%.

　i) Explain how a power rating of 18 W for the new design will give the same useful light output as the original bulb. Use calculations to support your answer. (3 marks)

　ii) Calculate the cost per week, as in part (a) for operating the new bulb with a power rating of 18 W. (2 marks)

　iii) Calculate how much money the new 18 W bulb would save per week. (1 mark)

c) Suggest whether you think the EU decision to ban 100 W bulbs and encourage the use of more energy efficient designs is the right thing to do. In your answer you should consider the environmental impact of energy saving measures. You should also include a piece of information that you would need to know to make a more informed decision. (6 marks)

Score / 14

Electricity Matters

Multiple-choice questions

Choose just one answer: A, B, C or D.

1 Which of the following statements is false? (1 mark)
- A a step-up transformer increases voltage
- B a step-up transformer decreases current
- C a step-up transformer increases current
- D a step-up transformer will only work with an alternating current

2 Which of the following statements is true? (1 mark)
- A transformers can change the power
- B transformers can do work
- C transformers are 100% efficient
- D transformers can change the value of an alternating current

3 Where is the voltage stepped-down in the UK National Grid? (1 mark)
- A power lines
- B pylons
- C power station
- D sub-station

4 Which of the following is NOT a disadvantage of nuclear power stations? (1 mark)
- A being at risk from terrorism
- B contribution to the greenhouse effect
- C accidental emission of radioactive waste
- D high decommissioning costs

5 Which of the following is NOT associated with nuclear power stations? (1 mark)
- A uranium
- B plutonium
- C carbon dioxide
- D radiation

Score / 5

Short-answer questions

1 a) Fill in the missing words to complete the information about the National Grid. (5 marks)

The National Grid is the network that distributes .... power around the country. It connects the power stations where electricity is .... to the users in homes and businesses. Electricity provided to the National Grid is generated in such a way that the current .... The frequency at which this happens is 50 Hz in the UK. The electricity reaches our homes via step-up transformers, ... (often carried by on pylons), and .... transformers.

b) State one advantage and one disadvantage of the National Grid.

(2

GCSE-style questions

Answer all parts of all questions. Continue on a separate sheet of paper if necessary.

1 In an experiment to investigate heat transfer, four different coloured (but otherwise identical) boxes of boiling water are allowed to cool for 3 minutes. The following results are obtained:

Colour of box	Final temp. (°C)
Silver	80
Matt black	60
White	75
Shiny black	70

a) What is the initial temperature of all the boxes? (1 mark)

b) Explain why it is important that the initial temperature of the water, the size and shape of the box and the amount of water in each box are the same. (2 marks)

c) Which box has... (2 marks)

  i) cooled the most? .... **ii)** cooled the least? ....

d) Explain these observations, referring in your answer to the possible contributions of conduction, convection and radiation as mechanisms of heat transfer. (4 marks)

e) Suggest another factor that you would consider in order to improve the reliability of the results. (1 mark)

Score / 10

How well did you do?

0–5 Try again | 6–11 Getting there | 12–17 Good work | 18–21 Excellent!

For more information on this topic, see pages 14–15 of your Success Revision Guide.

Describing Waves

Multiple-choice questions

Choose just one answer, A, B, C or D.

1 For which type of wave are the oscillations parallel to the direction in which the wave travels? (1 mark)
- A transverse
- B longitudinal
- C electromagnetic
- D water

2 Which of the following formulae correctly calculates the speed of a wave? (1 mark)
- A wavelength × frequency
- B frequency × amplitude
- C $\dfrac{\text{wavelength}}{\text{amplitude}}$
- D $\dfrac{\text{frequency}}{\text{wavelength}}$

3 What is the highest point on a transverse wave called? (1 mark)
- A compression
- B trough
- C amplitude
- D crest

4 Sound is an example of which type of wave? (1 mark)
- A longitudinal
- B seismic
- C electromagnetic
- D transverse

5 What is the lowest point on a transverse wave? (1 mark)
- A rarefaction point
- B trough
- C oscillation
- D crest

Score / 5

Short-answer questions

1 Calculate the wavelength of water waves of frequency 2 Hz travelling at 20 cm/s. (2 marks)

2 Calculate how far a sound wave of wavelength 34 cm and frequency 1000 Hz will travel in 5 s. (2 marks)

3 True or false? True False (3 marks)

- A Amplitude is measured from the top of the crest to the bottom of the trough.
- B The distance from one crest to the next crest is a wavelength.
- C Wavelength is the total length of a series of waves from the first crest to the last crest.

Score / 7

GCSE-style questions

Answer all parts of all questions. Continue on a separate sheet of paper if necessary.

1 Harry is listening to music on his MP3 player. The electrical signal from the player is fed to the headphones in order to produce the sound that Harry hears.

  a) Complete the following passage, inserting words into the spaces or choosing from the options provided. (4 marks)

  The sound that Harry hears is due to .... in the air produced by the headphones. Electrical .... is transferred to the air particles to create a sound wave. Sound is an example of a **transverse / longitudinal** wave, in which the air particles move in a direction **parallel / perpendicular** to the direction of wave travel.

  b) The diagram alongside shows a transverse wave, such as a water wave. It could also be taken to represent the electrical signal producing a steady tone in Harry's headphones.

  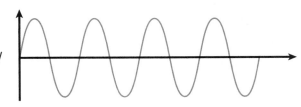

  On the diagram label the following features:
  i) amplitude **ii)** wavelength. (2 marks)

  c) i) State how the wave would differ if it was to show an electrical signal producing a quieter sound in the headphones. (1 mark)

  ....

  ii) State how the wave would differ if it was to show an electrical signal producing a sound of higher pitch. (1 mark)

  ....

  d) Explain why the diagram is not a good representation of the sound wave that is produced by the headphones. (3 marks)

  ....

  ....

  ....

  e) Although the wave shown does not directly represent sound it can nevertheless be a useful model if it is used to show how the air pressure rather than displacement changes.

  Now complete the following continuation of the paragraph in part **a)**. (3 marks)

  The sound wave consists of alternating high and low pressure regions. Areas of high pressure are known as **compressions / rarefactions**. Areas of low pressure are called **compressions / rarefactions**. In a sound of lower pitch the compressions occur **closer together / further apart**. In a sound of greater volume the **amplitude / frequency** is higher.

  Score / 14

How well did you do?

| 0–6 | Try again | 7–12 | Getting there | 13–19 | Good work | 20–26 | Excellent! |

For more information on this topic, see pages 18–19 of your Success Revision Guide.

Wave Behaviour

Multiple-choice questions

Choose just one answer: A, B, C or D.

1 What is the correct term for the change of direction, that accompanies a change in speed, when a wave travels from one medium to another? **(1 mark)**
- A reflection
- B refraction
- C diffraction
- D dispersion

2 Which term refers to the change in direction, without a change in speed, when a wave hits a suitable surface? **(1 mark)**
- A refraction
- B incidence
- C reflection
- D an echo

3 When a wave strikes a surface, the angle between the wave's direction of travel and the normal is called the angle of: **(1 mark)**
- A incidence
- B approach
- C refraction
- D reflection

4 Which colour of light is refracted the most by a prism? **(1 mark)**
- A red
- B green
- C blue
- D violet

5 Which term means 'the spreading of waves when going through a narrow gap'? **(1 mark)**
- A refraction
- B diffraction
- C reflection
- D dispersion

Score / 5

Short-answer questions

1 The following ray diagram showing the reflection of a beam of light contains an error.

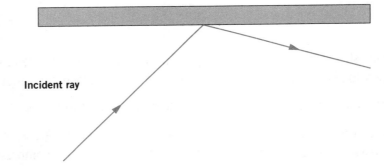

Incident ray

a) Circle the part of the diagram with the error. (1 mark)

b) Draw on the diagram to show how the error should be corrected. (1 mark)

c) Complete the following statement. (1 mark)

When a wave is reflected, the angle of reflection...

Score / 3

GCSE-style questions

Answer all parts of all questions. Continue on a separate sheet of paper if necessary.

1 When you look into a body of water from above, it is very difficult to accurately judge distances beneath the surface of the water.

   a) Explain why the bottom of a swimming pool appears to be closer, i.e. the pool seems shallower, than it really is. Draw a ray diagram to support your answer. **(3 marks)**

   b) Describe why the same difficulties would not be faced by an observer who is underwater. **(2 marks)**

2 Colin is walking down the road towards a T-junction. He can hear the sound of approaching sirens, but he only sees the police car when it speeds by the end of the road.

Explain why it is possible to hear sounds but not see around the corner. **(3 marks)**

3 When sound waves hit a surface they can be reflected. In everyday language, this is known as an echo. A battleship floating on the surface of the ocean makes use of this behaviour by using a sonar detector to locate underwater objects, such as submarines. It does this by recording the time taken to 'hear' the echo. If ultrasonic waves at a frequency of 30 kHz travel through the water at 1500 m/s and the detector receives the signals after a delay of 0.5 s, calculate the following:

   a) the wavelength of the sonar waves. **(2 marks)**

   b) the distance to the submarine. **(2 marks)**

Score / 12

How well did you do?

| 0–5 | Try again | 6–11 | Getting there | 12–16 | Good work | 17–20 | Excellent! |

For more information on this topic, see pages 20–21 of your Success Revision Guide.

Seismic Waves and the Earth

Multiple-choice questions

Choose just one answer, A, B, C or D.

1 Which seismic waves travel fastest? (1 mark)
 A earthquake waves
 B S-waves
 C P-waves
 D tectonic waves

2 Which seismic waves are longitudinal? (1 mark)
 A P-waves
 B earthquake waves
 C S-waves
 D tectonic waves

3 S-waves waves are: (1 mark)
 A longitudinal
 B transverse
 C primary
 D sound

4 What is the name for the outermost layer of the Earth? (1 mark)
 A core
 B magma
 C crust
 D mantle

5 In which layer of the Earth are the rocks semi-molten? (1 mark)
 A magma
 B crust
 C core
 D mantle

Score / 5

Short-answer questions

1 Fill in the missing words to explain the behaviour of P-waves and S-waves. (7 marks)

When an earthquake occurs, waves spread from the site of the quake throughout the Earth. These waves are called .... waves, of which there are two types: P-waves and S-waves. The .... -waves travel fastest and are the first to be detected. They are pressure waves that are capable of travelling through both solid and .... materials. The .... -waves are detected later because they are unable to travel through .... parts of the Earth. Monitoring this wave activity using pieces of equipment called .... allows scientists to analyse the Earth's inner structure.

2 State which currents in the mantle of the Earth cause the continental tectonic plates to move. (1 mark)

....

3 State two pieces of evidence that support Wegener's theory of continental drift. (2 marks)

....
....
....

Score / 10

GCSE-style questions

Answer all parts of all questions. Continue on a separate sheet of paper if necessary.

1 A scientist in an earthquake monitoring station obtains the following reading on a seismometer.

a) On the seismograph, label
   i) the P-waves ii) the S-waves. (2 marks)

b) The scientist uses the following formula to estimate the distance of the earthquake from the monitoring station:

Distance from station (km) =

(Time for S-waves to arrive (s) – Time for P-waves to arrive (s)) x 8

Use the seismograph and formula to estimate the distance of the earthquake from the monitoring station. (2 marks)

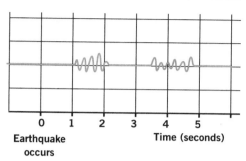

c) For the same earthquake, some other stations only picked up readings for one set of waves on their seismometers.

   i) Describe where these stations are likely to be located in relation to the epicentre of the earthquake. (1 mark)

   ii) What is the name given to the region in which these stations are located?
   Tick the correct option.

   A Safe haven
   B Shadow zone
   C Refraction region
   D Seismic area

d) Explain how understanding how P-waves and S-waves behave has also helped scientists to develop a better understanding of the structure of the Earth. (3 marks)

Score / 8

How well did you do?

| 0–6 Try again | 7–12 Getting there | 13–17 Good work | 18–23 Excellent! |

For more information on this topic, see pages 22–23 of your Success Revision Guide.

The Electromagnetic Spectrum

Multiple-choice questions

Choose just one answer: A, B, C or D.

1 Which is the longest wavelength region of the electromagnetic spectrum? **(1 mark)**
 A light
 B radio
 C X-ray
 D infrared

2 Which part of the spectrum has a wavelength slightly greater than that of visible light? **(1 mark)**
 A X-ray
 B microwave
 C light
 D infrared

3 Which of the following wavelengths falls into the spectrum of visible light? **(1 mark)**
 A 3×10^2
 B 4×10^{-2}
 C 5×10^{-7}
 D 6×10^{-12}

4 Which of these is NOT a possible effect of radiation absorption? **(1 mark)**
 A the material heats up
 B atoms in the material are ionised
 C chemical reactions are more likely to occur
 D the material cools down

5 Which electromagnetic waves are sometimes known as heat radiation? **(1 mark)**
 A light
 B radio
 C X-ray
 D infrared

Score / 5

Short-answer questions

1 a) Underline the correct words to complete this description of the electromagnetic spectrum. **(3 marks)**

The electromagnetic spectrum is **continuous / intermittent** from the longest to shortest wavelengths.

All electromagnetic waves are **longitudinal / transverse** and travel through space at a speed of **300 / 300 000 / 300 000 000** m/s.

b) Put the following regions of the electromagnetic spectrum in the correct order, starting with the longest wavelength. **(7 marks)**

radio waves, X-rays, visible light, gamma rays, microwaves, infrared, ultraviolet

....

2 Many conventional diagrams showing the electromagnetic spectrum can be misleading when it comes to the shortest wavelengths, i.e. gamma rays and X-rays.

Explain why this is. **(2 marks)**

....

....

Score / 12

GCSE-style questions

Answer all parts of all questions. Continue on a separate sheet of paper if necessary.

1 Radiation from all parts of the electromagnetic spectrum travels through space at the 'speed of light' which is 300 000 km/s.

  a) Calculate the frequency of microwaves of wavelength 2 cm. (2 marks)

  b) Calculate the wavelength of ultraviolet light of frequency 6×10^{15} Hz. (2 marks)

  c) The following data is found on a radio-mast: 'Range of frequencies is 300 kHZ to 500 MHz'.

  i) Calculate the corresponding range of wavelengths. (2 marks)

  ii) The optimum length for a mast is usually one quarter of a wavelength. State how long you would make the mast for optimum performance. (1 mark)

2 a) Describe what is meant by 'intensity' in reference to radiation striking a surface. (1 mark)

  b) State two ways in which intensity can be increased. (2 marks)

  c) The energy in electromagnetic radiation is arranged in 'packets' called photons. The photon energy is directly proportional to the frequency.

  In a photographic dark room, the photographer processes film that is extremely light sensitive. With reference to photon energy, explain why red lights can be used in dark rooms but other coloured lights cannot. (3 marks)

Score / 13

How well did you do?

| 0–8 Try again | 9–14 Getting there | 15–22 Good work | 23–30 Excellent! |

For more information on this topic, see pages 26–27 of your Success Revision Guide.

Light, Radio Waves and Microwaves

Multiple-choice questions

Choose just one answer: A, B, C or D.

1. Which of the following radiations does NOT reach the Earth's surface from the Sun? **(1 mark)**
   A microwaves   B high energy infrared
   C visible light   D low energy ultraviolet

2. Which of the following is NOT a property of radio waves and microwaves? **(1 mark)**
   A they are reflected by metal surfaces
   B they can have a heating effect
   C they have an ionising effect
   D they have much greater wavelengths than visible light

3. Medium wave radio signals travel long distances by reflection from a layer of charged particles called the: **(1 mark)**
   A atmosphere   B stratosphere
   C heliosphere   D ionosphere

4. Microwaves can be used for cooking. What is the purpose of the metal grid on the door of a microwave oven? **(1 mark)**
   A to help heat the food
   B to increase the microwave intensity
   C to prevent microwaves from getting out of the oven
   D to produce the microwaves

5. Plants rely on the energy from visible light for: **(1 mark)**
   A respiration   B photosynthesis
   C digestion   D transmission

Score / 5

Short-answer questions

1. When a radio signal is broadcast, an alternating current of electrons in the transmitter aerial causes the radio wave to be emitted. Describe what happens to the electrons in the receiver aerial when the radio signal is detected. **(1 mark)**

2. a) State why the casing of a microwave oven has to be opaque to the waves. **(1 mark)**

   b) Explain why a microwave oven is good for cooking a ready meal, but not for making toast. **(2 marks)**

3. Fill in the missing words to complete the following passage. **(3 marks)**

   Detection of light by a camera is similar to how the eye detects light. In both, light passes through a small hole to produce a small, upside down .... A .... is used to gather and focus more light. The photographic film or digital screen inside the camera plays the role of the .... in the eye.

Score / 7

GCSE-style questions

Answer all parts of all questions. Continue on a separate sheet of paper if necessary.

1 Microwaves are absorbed by water. The frequency of the microwaves determines how well they are absorbed.

a) With reference to the statement above, explain how microwaves are used to cook food in microwave ovens. **(6 marks)**

b) Explain why the property of microwaves described above means that only certain frequencies of microwave can be used for wireless communications. **(2 marks)**

c) Using your knowledge of the properties of microwaves, explain how microwaves might be used to help with weather forecasting. **(3 marks)**

2 Electromagnetic radiation from the Sun that reaches the Earth covers a broader spectrum than the visible region that we see.

a) Apart from visible light, state two types of radiation from the Sun that penetrate to the Earth's surface. **(2 marks)**

b) State one type of radiation from the Sun that fails to reach the surface and explain why. **(2 marks)**

Score / 15

How well did you do?

0–6 Try again | 7–12 Getting there | 13–19 Good work | 20–27 Excellent!

For more information on this topic, see pages 28–29 of your Success Revision Guide.

Wireless Communications 1

Multiple-choice questions

Choose just one answer: A, B, C or D.

1 Which waves have to be used in direct 'line of sight' telecommunication? **(1 mark)**
  A analogue
  B digital
  C radio waves
  D microwaves

2 Which waves can pass through the Earth's atmosphere and diffract around large obstacles? **(1 mark)**
  A analogue
  B digital
  C radio waves
  D microwaves

3 Which signals are coded by 1s and 0s? **(1 mark)**
  A analogue
  B digital
  C radio waves
  D microwaves

4 Which signals can have a continuous range of values? **(1 mark)**
  A analogue
  B digital
  C radio waves
  D microwaves

5 Which signals give the best quality of reception and sound reproduction? **(1 mark)**
  A analogue
  B digital
  C radio waves
  D microwaves

Score / 5

Short-answer questions

1 Draw one line from each word to the correct definition. (3 marks)

Regeneration	The changing of a signal that codes the information being sent
Amplification	The process of cleaning up a digital signal
Noise	Unwanted signals
Modulation	Increasing the size of a signal

2 True or false? True False (6 marks)

  A We can hear radio waves but we cannot hear microwaves. ☐ ☐
  B Microwaves pass through the Earth's atmosphere. ☐ ☐
  C Radio waves have a lower frequency than microwaves. ☐ ☐
  D Radio waves diffract more than microwaves. ☐ ☐
  E Light travels faster than radio waves. ☐ ☐
  F Radio waves and microwaves can be used as carrier waves. ☐ ☐

Score / 9

GCSE-style questions

Answer all parts of all questions. Continue on a separate sheet of paper if necessary.

1 Microwaves and radio waves are both used extensively for telecommunication.

Microwave transmitters and receivers must be correctly aligned, whereas this is not the case for radio or television reception.

With reference to their wavelengths, explain this behaviour of radio waves and microwaves. **(5 marks)**

2 Digital television broadcasting is replacing the old analogue system.
   a) Number the statements **A–E** below to describe how information is transmitted using digital technology. **(5 marks)**

   A The codes are added to a carrier wave.

   B The information is divided into tiny segments of time.

   C The signal is received and decoded to recover the original information.

   D Each segment is given a numerical value using binary code.

   E The wave is transmitted.

   b) State what a binary code is. **(1 mark)**

   c) An advertisement for the national switchover to digital transmission claims 'unwanted noise on your television screen is a thing of the past with digital television'.

   i) Describe what is meant by the term 'noise' in this context. **(2 marks)**

   ii) ✎ Explain whether or not the claim being made in the advertisement is justified. **(6 marks)**
   (Answer on a separate piece of paper.)

Score / 19

How well did you do?

0–10 Try again | 11–18 Getting there | 19–25 Good work | 26–33 Excellent!

For more information on this topic, see pages 30–31 of your Success Revision Guide.

Wireless Communications 2

Multiple-choice questions

Choose just one answer: A, B, C or D.

1. Which electromagnetic waves are used for satellite communications? (1 mark)
   - A radio waves
   - B microwaves
   - C light waves
   - D heat waves

2. Which electromagnetic waves are used for television broadcasts? (1 mark)
   - A radio waves
   - B light waves
   - C microwaves
   - D sound waves

3. Which of the following is NOT an advantage of a wireless phone connection? (1 mark)
   - A no wires are needed
   - B the user needs to be close to a transmitter mast
   - C increased portability
   - D phone and internet connectivity are constantly available

4. When carrying out investigations into the health effects of mobile phone use, why do scientists match their samples so that they are comparing the same type of people? (1 mark)
   - A to minimise the number of external factors that could affect the results
   - B to minimise the cost of the investigation
   - C because they are only interested in the effects on certain groups of people
   - D because it makes it easier to find people to study

Score / 4

Short-answer questions

1. Computer and communication technology is increasingly 'wireless'.

   a) State two pieces of equipment that a personal computer might routinely be connected to. (2 marks)

   b) State three advantages of using wireless communication in the workplace or at home. (3 marks)

2. Match the graphs A to C with the most appropriate description of how the variables relate. (2 marks)
   1. negative correlation between X and Y
   2. no correlation between X and Y
   3. positive correlation between X and Y

Score / 7

GCSE-style questions

Answer all parts of all questions. Continue on a separate sheet of paper if necessary.

1 The 'explosion' in recent years of the use of microwave technology for mobile phones has caused some people to be alarmed at the possible health implications.
One concern is that because the handset is held close to the head, the microwaves have a heating effect on the brain.

   a) State which other use of microwaves would seem to support the possibility of such a risk. **(1 mark)**

   b) It is generally thought that the risk to health caused by the heating effect of microwaves from mobile phones is negligible. With reference to your answer to part **a)**, explain why this is. **(2 marks)**

   c) The only reliable way to determine if there is a significant health risk is to carry out a long-term study. Describe why it is necessary for the study to run for many years. **(2 marks)**

   d) State two other factors that would need to be considered when setting up such a study to ensure reliable results. **(2 marks)**

   e) Describe why is it important that the scientists who carry out such investigations are independent of the mobile phone companies. **(1 mark)**

   f) ✎ A mobile phone company offers to pay a school money to install a transmitter mast on the top of a school building. Many of the parents and teachers are planning a protest.

   Suggest whether you think the proposal is a good idea or not and explain your answer. **(6 marks)**

Score / 14

How well did you do?

| 0–6 Try again | 7–13 Getting there | 14–19 Good work | 20–25 Excellent! |

For more information on this topic, see pages 32–33 of your Success Revision Guide.

Infrared

Multiple-choice questions

Choose just one answer: A, B, C or D.

1 Optical fibres can carry signals to and from many homes and businesses at the same time, because of a process called: **(1 mark)**
 A multiplexing
 B time sharing
 C divergence
 D total internal reflection

2 Which of the following applications does not use infrared? **(1 mark)**
 A security sensors
 B remote control devices
 C wireless Internet access
 D thermal imaging

3 The process by which light is transmitted using optical fibre requires the angle of incidence to be: **(1 mark)**
 A less than 90°
 B 0°
 C greater than the critical angle
 D less than the critical angle

4 What is the name of the process by which infrared radiation and light are transmitted down optical fibres? **(1 mark)**
 A total external reflection
 B total internal reflection
 C diffraction
 D refraction

5 A device that produces a narrow beam of intense radiation of one colour is called a: **(1 mark)**
 A laser
 B Sun lamp
 C LED
 D bulb

Score / 5

Short-answer questions

1 Describe how it is possible to make a piece of metal emit visible light. **(1 mark)**

....

2 True or false? True False **(5 marks)**
 A Light is a longitudinal wave. ☐ ☐
 B Red light has a higher frequency than blue light. ☐ ☐
 C X-ray photons have more energy than ultraviolet photons. ☐ ☐
 D Radio waves diffract more easily than microwaves. ☐ ☐
 E Lasers are used to read the information stored on a compact disc. ☐ ☐

3 Fill in the missing words to complete the following passage.

Digital signals can be efficiently transmitted as infrared along optical fibres made of two layers of

.... Each layer has a different .... Every time the light

beam hits the boundary between the two layers .... reflection directs

it back into the fibre.

(3 marks)

Score / 9

GCSE-style questions

Answer all parts of all questions. Continue on a separate sheet of paper if necessary.

1 Originally Broadband Internet was delivered along telephone lines, but the use of new fibre optic cables is becoming increasingly popular.

   a) State which type of electromagnetic waves are used to transmit signals down fibre optic cables. **(1 mark)**

   b) i) Describe how the frequency of this type of wave compares to radio waves and microwaves. **(1 mark)**

   ii) Explain the advantages of using fibre optic cables to transmit Broadband. **(3 marks)**

2 A naturalist wishing to study the behaviour of a nocturnal mammal invests in a camera with night vision technology, which is capable of detecting infrared.

   a) Explain why such technology will be useful to monitor the activity of nocturnal animals. **(2 marks)**

   b) The camera manual states that 'The detected infrared is converted and shown in false colour'. Describe what this means. **(2 marks)**

   c) Similar technology is employed by rescue services, for example, to seek out survivors buried under rubble, following an earthquake. In this context, the use of infrared it is often referred to as 'thermal imaging'. Explain what this means and what limitations the technique has. **(4 marks)**

Score / 13

How well did you do?

| 0–6 | Try again | 7–13 | Getting there | 14–20 | Good work | 21–27 | Excellent! |

For more information on this topic, see pages 34–35 of your Success Revision Guide.

The Ionising Radiations

Multiple-choice questions

Choose just one answer: A, B, C or D.

1 Which electromagnetic waves come from radioactive sources? (1 mark)
- A UV
- B X-ray
- C gamma
- D light

2 Which of the following electromagnetic waves are the most highly penetrating? (1 mark)
- A gamma
- B light
- C UV
- D radio

3 Which electromagnetic waves can cause sun burn? (1 mark)
- A light
- B UV
- C X-ray
- D gamma

4 Which waves pose the least risk of cell damage due to ionisation? (1 mark)
- A gamma
- B X-ray
- C UV
- D light

Score / 4

Short-answer questions

1 Describe one way in which UV light can be used to prevent / detect criminal activity. (1 mark)

2 a) Describe how radiation can ionise atoms. (1 mark)

b) State the three types of ionising radiation released by radioactive materials. (3 marks)

c) i) State which type of ionising radiation is the most penetrating. (1 mark)

ii) State which type of ionising radiation is the least penetrating. (1 mark)

d) State which two types of ionising radiation are particles and describe what they consist of. (2 marks)

Score / 9

GCSE-style questions

Answer all parts of all questions. Continue on a separate sheet of paper if necessary.

1 X-rays are useful in diagnosing broken bones.

a) Explain why the radiographer leaves the room while the X-ray of a patient is taken. **(2 marks)**

b) Using the idea of risk versus benefit, explain why it is acceptable to expose the patient to X-rays. **(3 marks)**

2 A scientist is asked to carry out an investigation into whether there is a positive correlation between the level of sun bed use and an individual's chances of developing skin cancer.

a) Explain why exposure to ionising radiation is likely to increase the risk of cancer. **(4 marks)**

b) Describe what type of sample of people the scientist should aim to use for this investigation. **(2 marks)**

c) Explain how the scientist should group the people in his sample. **(3 marks)**

d) The scientists decide to us a control group. Explain what this means. **(2 marks)**

e) Explain why an investigation into correlation can never be totally conclusive. **(2 marks)**

Score / 18

How well did you do?

| 0–8 | Try again | 9–16 | Getting there | 17–24 | Good work | 25–31 | Excellent! |

For more information on this topic, see pages 36–37 of your Success Revision Guide.

The Atmosphere

Multiple-choice questions

Choose just one answer: A, B, C or D.

1 Infrared frequencies, emitted by the Earth, are absorbed by gases in the atmosphere. Which of the following is NOT one of those gases? (1 mark)
- A methane
- B carbon dioxide
- C water vapour
- D ozone

2 Which of the following does NOT contribute to the greenhouse effect? (1 mark)
- A burning fossil fuels
- B cattle farming
- C deforestation
- D rising sea levels

3 Which of these is produced when burning fossil fuels? (1 mark)
- A oxygen
- B carbon dioxide
- C hydrogen
- D ozone

4 The ozone layer is responsible for: (1 mark)
- A absorbing dangerous UV radiation
- B filtering the air we breathe
- C filtering out infrared radiation
- D absorbing carbon dioxide

5 Depletion of which of these gases in the atmosphere is likely to lead to more cases of skin cancer? (1 mark)
- A ozone
- B carbon dioxide
- C water vapour
- D methane

Score / 5

Short-answer questions

1 The 'greenhouse effect' is so called because it mirrors the way in which a greenhouse works.

a) State which type of electromagnetic radiation is absorbed by the glass in a greenhouse. (1 mark)

b) State which type of electromagnetic radiation passes though the glass. (1 mark)

2 True or false? True False (6 marks)

- A Infrared radiation can cause skin cancer. ☐ ☐
- B Low frequency ultraviolet radiation cannot penetrate the atmosphere. ☐ ☐
- C CFCs are thought to affect the ozone layer in the Earth's atmosphere. ☐ ☐
- D Infrared photons are less energetic than ultraviolet photons. ☐ ☐
- E If the distance to a source of light is double the intensity received halves. ☐ ☐
- F The intensity of sunlight is at its daily maximum around noon. ☐ ☐

Score / 8

GCSE-style questions

Answer all parts of all questions. Continue on a separate sheet of paper if necessary.

1 Global warming is a major environmental concern. One of the main contributing factors is thought to be the increase in emissions of 'greenhouse gases' over the past 200 years.

   a) State two consequences of global warming. **(2 marks)**

   b) Explain what is meant by the term 'greenhouse gases'. **(2 marks)**

   c) Carbon dioxide is a greenhouse gas. Describe two reasons for increased carbon dioxide emissions over the last 200 years. **(2 marks)**

   d) Some scientists dispute that global warming is taking place. Give one possible reason for this. **(1 mark)**

2 In recent years, as a result of international agreement, CFCs have been replaced by other chemicals in aerosol sprays and refrigerants. This is because the use of CFCs has been linked to a hole in the ozone layer.

   a) Explain what the ozone layer is and why it is important. **(2 marks)**

   b) Describe one of the likely consequences to humans of damage to the ozone layer. **(2 marks)**

   c) Explain how CFCs are thought to have caused a hole in the ozone layer. **(2 marks)**

Score / 13

How well did you do?

0–8 Try again | 9–14 Getting there | 15–20 Good work | 21–26 Excellent!

For more information on this topic, see pages 38–39 of your Success Revision Guide.

The Solar System

Multiple-choice questions

Choose just one answer: A, B, C or D.

1 Which of the following planets is furthest from Earth? (1 mark)
 A Jupiter
 B Mercury
 C Neptune
 D Mars

2 Which is the only star in the Solar System? (1 mark)
 A Moon
 B Earth
 C Sun
 D Halley's comet

3 Approximately how old is the Solar System thought to be? (1 mark)
 A 5 000 000 000 000 years old
 B 5 000 000 000 years old
 C 5 000 000 years old
 D 5 000 years old

4 Which is the biggest planet in the Solar System? (1 mark)
 A Jupiter
 B Saturn
 C Neptune
 D Pluto

5 Which of the following planets is closest to the Sun? (1 mark)
 A Jupiter
 B Mercury
 C Saturn
 D Neptune

Score / 5

Short-answer questions

1 a) Describe how the 'inner' planets differ from the 'outer' planets. (2 marks)

b) State what lies between the inner and outer planets. (1 mark)

2 Draw one line from each type of body found in the Solar System to the most accurate description. (4 marks)

Sun	A natural satellite in orbit around a planet
Moon	A small rock, usually from a comet, entering the Earth's atmosphere
Asteroid	The star at the centre of the Solar System
Comet	A rock, up to about 1 km in diameter, in orbit around the Sun
Meteor	Accumulated ice and dust, with a nucleus that vaporises when it comes close the Sun

Score / 7

GCSE-style questions

Answer all parts of all questions. Continue on a separate sheet of paper if necessary.

1 The following data are for the orbital periods and surface temperature of planets in the Solar System.

	Distance from Sun (1×10^6 km)	Orbital period	Average surface temperature (°C) (on side facing Sun)
Mercury	60	88 days	430
Venus	110	220 days	470
Earth	150	365 1/4 days	20
Mars	230	2 years	−20
Jupiter	780	12 years	−150
Saturn	1400	30 years	−180
Uranus	2900	84 years	−210
Neptune	4500	160 years	−220

a) On the grid, plot a graph for average surface temperature against distance from Sun. **(4 marks)**

b) Explain the relationship between surface temperature and orbital distance. **(2 marks)**

c) State which planet is anomalous for this graph **(1 mark)**

d) Suggest one reason for this anomaly. **(1 mark)**

Score / 8

How well did you do?

0–5 Try again 6–10 Getting there 11–16 Good work 17–20 Excellent!

For more information on this topic, see pages 42–43 of your Success Revision Guide.

Space Exploration

Multiple-choice questions

Choose just one answer: A, B, C or D.

1 Which of these radiations does not pass through the Earth's atmosphere? **(1 mark)**
A light
B X-rays
C radio waves
D microwaves

2 Which of these explains why telescopes are put into orbit on satellites? **(1 mark)**
A it is easier
B it saves money
C it is safer
D it gives better images

3 Which of these is a reason for manned space travel? **(1 mark)**
A it is dangerous
B unplanned adjustments and repairs can be made during the mission
C it takes a long time
D it costs money

4 Which of these organisations is concerned with space travel? **(1 mark)**
A NASA
B NATO
C UN
D UNESCO

5 Which of the following is NOT evidence that the Moon was formed during the collision of a planet with Earth? **(1 mark)**
A rock samples from the Moon match rock samples from Earth
B the Moon has a relatively small iron core unlike other planets
C there are no oceans on the Moon
D the Moon has igneous rocks, although there has been no recent volcanic activity

Score / 5

Short-answer questions

1 The current theory of how the Moon was formed is sometimes called the 'Giant Impact Theory'. Ring the correct options to complete the passage and describe this theory. **(4 marks)**

The Giant Impact Theory states that early on in its history, the **Earth / Sun / Mars** collided with another object, about the size of **Jupiter / Pluto / Mars**.

The impact destroyed the other body and propelled debris into **orbit / outer space / the Sun**.

Over time, the debris **dispersed / compacted / imploded** to form the Moon.

2 Telescopes have been built to explore all regions of the electromagnetic spectrum. Draw lines to link the region of the electromagnetic spectrum with the optimal location for a telescope. **(4 marks)**

X-ray
Radio wave
Microwave
Infrared

Ground-based
In orbit above the Earth's atmosphere

Score / 8

GCSE-style questions

Answer all parts of all questions. Continue on a separate sheet of paper if necessary.

1 In 2010, US President Barack Obama made this comment: 'By the mid-2030s, I believe we can send humans to orbit Mars and return them safely to Earth.'

a) NASA has successfully used unmanned robotic Rovers to explore Mars. The Rovers are controlled by engineers, who upload a series of commands each day.

   i) Describe why this method of control is used instead of traditional radio control methods. **(2 marks)**

   ii) Describe one disadvantage of operating Rovers by uploading commands each day. **(1 mark)**

b) State one advantage of a manned mission to Mars over an unmanned mission. **(1 mark)**

c) Explain why it takes careful planning before a manned mission can be undertaken. Your answer should make reference to four specific issues that need to be considered. **(4 marks)**

2 The Hubble telescope, in geostationary orbit, has produced images of significantly greater detail than those produced from the most sophisticated ground-based optical telescopes. Describe the advantages and disadvantages of orbiting telescopes compared to ground-based telescopes. **(4 marks)**

Score / 12

How well did you do?

| 0–6 Try again | 7–12 Getting there | 13–18 Good work | 19–25 Excellent! |

For more information on this topic, see pages 44–45 of your Success Revision Guide.

A Sense of Scale

Multiple-choice questions

Choose just one answer: **A, B, C or D.**

1 At what speed does light travel through space? **(1 mark)**
- A 3000 km/s
- B 300 000 km/s
- C 300 km/s
- D 3 000 000 km/s

2 Which of these is NOT a reason why one star might look brighter than another? **(1 mark)**
- A it is moving faster than the other star
- B it is bigger than the other star
- C it is hotter than the other star
- D it is closer to Earth than the other star

3 Which of the following does NOT orbit the Sun? **(1 mark)**
- A Jupiter
- B Halley's comet
- C the Milky Way
- D the asteroid Vesta

4 Which of these bodies is the smallest? **(1 mark)**
- A Earth
- B asteroids
- C dwarf planets
- D Moon

5 The galaxy in which we live is called: **(1 mark)**
- A the Andromeda Galaxy
- B the Solar System
- C the asteroid belt
- D the Milky Way

Score / 5

Short-answer questions

1 True or false? True False **(5 marks)**

- A A light year is a unit of time.
- B It takes approximately 8 minutes for light from the Sun to reach the Earth.
- C Blue stars are hotter than red stars.
- D A bright star **must** be hotter than a dim star.
- E We can use parallax to judge the distance away an object is.

2 Put the following in order of increasing size: **(4 marks)**

- A The distance from the Earth to the Sun
- B The distance from the Earth to the Moon
- C The estimated diameter of our galaxy
- D 1 light year
- E The distance to the 'nearby' star Alpha Centauri

Score / 9

GCSE-style questions

Answer all parts of all questions. Continue on a separate sheet of paper if necessary.

1 An astronomer discovers a new star using a telescope, when observing a lesser studied region of the sky.

a) Over the next year he uses parallax to determine the star's distance from Earth.

i) Explain what is meant by 'parallax' and how this can be used to judge the distance to nearby stars. **(2 marks)**

ii) Describe why measuring distance using the parallax method requires observation over a long period of time. **(2 marks)**

b) A number of factors affect how bright a star appears. State two such factors. **(2 marks)**

c) When an iron rod is heated up it starts to glow, initially red then tending towards white. This is why we talk about things being 'red hot' and 'white hot'. Use this analogy to explain how the colour of the light from a star gives information about its surface temperature. **(3 marks)**

d) Very large distances cannot be measured using parallax. Describe how can we judge the distance to stars which are many millions of light years away. **(1 mark)**

e) 'The astronomer is not observing the star in real time; rather, he sees it as it used to be.' Explain how this statement can be true. **(2 marks)**

Score / 12

How well did you do?

| 0–6 Try again | 7–12 Getting there | 13–19 Good work | 20–26 Excellent! |

For more information on this topic, see pages 46–47 of your Success Revision Guide.

Stars

Multiple-choice questions

Choose just one answer: A, B, C or D.

1 In which part of a star does thermonuclear fusion take place? (1 mark)
A core
B protostar
C nucleus
D nebula

2 Which gas is produced by the nuclear fusion of hydrogen? (1 mark)
A oxygen
B carbon dioxide
C hydrogen oxide
D helium

3 What is the explosion that occurs when a very massive star comes to an end? (1 mark)
A big bang
B supernova
C white dwarf
D red supergiant

4 What will our Sun become at the end of its 'life cycle'? (1 mark)
A red giant
B supernova
C black dwarf
D planetary nebula

5 Which is the second most common element in the Universe? (1 mark)
A hydrogen
B carbon
C oxygen
D helium

Score / 5

Short-answer questions

1 a) State which force is responsible for the formation of stars. (1 mark)

b) Describe why stars do not keep contracting because of this force. (2 marks)

2 Label this diagram of the life cycle of a star that is more massive than our Sun. (5 marks)

A   B   C   D   E

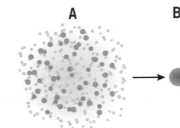
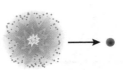

Score / 8

GCSE-style questions

Answer all parts of all questions. Continue on a separate sheet of paper if necessary.

1 After 10 000 million years the Sun is expected to run out of the hydrogen that fuels nuclear fusion.

a) Explain the stages involved when a star like the Sun comes towards the end of its 'life cycle'. **(6 marks)**

b) In contrast to the Sun, the star Rigel is expected to undergo a different fate. Rigel, in the constellation Orion, has a mass 17 times greater than that of the Sun. Describe how Rigel's fate will differ from that of the Sun. **(3 marks)**

2 Explain why the core of a star must be at a temperature of many millions of degrees Celsius. **(3 marks)**

3 In 1604, Kepler observed a supernova in the constellation Ophiuchus, within the Milky Way galaxy. It occurred at a distance of 20 000 light years.

Calculate the year in which the supernova actually happened. **(1 mark)**

Score / 13

How well did you do?

| 0–6 Try again | 7–12 Getting there | 13–19 Good work | 20–26 Excellent! |

For more information on this topic, see pages 48–49 of your Success Revision Guide.

Galaxies and Red-Shift

Multiple-choice questions

Choose just one answer: A, B, C or D.

1 What is the name of our galaxy? **(1 mark)**
- A Hubble
- B Milky Way
- C Halley
- D Andromeda

2 Who discovered the red-shift of distant galaxies? **(1 mark)**
- A Curtis
- B Doppler
- C Hubble
- D Shapley

3 What 'colour shift' would we expect for an object moving towards us very quickly? **(1 mark)**
- A red shift, due to increased apparent wavelength
- B blue shift, due to reduced apparent wavelength
- C red shift, due to reduced apparent wavelength
- D blue shift, due to increased apparent wavelength

4 Estimating the speed of a star relies on: **(1 mark)**
- A determination of the star's brightness
- B analysis of the star's emission spectrum
- C determination of the star's colour
- D analysis of the star's absorption spectrum

5 What is the name given to the effect which leads to the changing pitch of a moving police car siren? **(1 mark)**
- A Shapley Effect
- B Hubble Effect
- C Doppler Effect
- D Andromeda Effect

Score / 5

Short-answer questions

1 a) Describe how the speed of a galaxy moving away from us changes with distance. (1 mark)

b) Describe how the effect described in part **a)** is useful to astronomers who discover distant galaxies. (1 mark)

2 Fill in the blank spaces using the terms in the box below. (7 marks)

| bigger | closer | Doppler effect | blue | faster | Big Bang |
| slower | expanding | away | smaller | green | further |

Observation of light from distant galaxies shows that the universe is ....

This provides evidence for the .... theory. The .... the galaxy, the .... the red-shift, so the .... the galaxy is moving ....

This is an example of the .... If a galaxy were moving towards us, the light would be .... -shifted.

Score / 9

Beyond the Earth

44

GCSE-style questions

Answer all parts of all questions. Continue on a separate sheet of paper if necessary.

1 In the 1920s there was a great debate among astronomers about whether the Milky Way was one of many galaxies or the only galaxy.

   a) Explain why scientists sometimes disagree about an explanation although they have seen the same data. **(2 marks)**

   b) When further data become available it is often possible to eliminate some explanations and refine others.

   Describe three processes undertaken by the scientific community to evaluate such data. **(3 marks)**

   c) In the 1920s Harlow Shapley argued that the Milky Way was the only galaxy and that spiral nebulae were gas clouds. Heber Curtis argued that spiral nebulae were distant galaxies.

   Tick the statements below that are true and that show how Edwin Hubble produced evidence to support Curtis's theory. **(1 mark)**

   A He used a new telescope to observe the spiral nebulae. ☐
   B He used red-shift to help measure distances. ☐
   C He argued that you would not be able to see spiral nebulae if they were gas clouds. ☐
   D He proved that the spiral nebulae contained solid matter. ☐
   E He showed that spiral nebulae were outside the Milky Way and, therefore, could be other galaxies. ☐

2 ✏ Explain how the spectrum of light from distant stars enables us to identify the elements present in them and determine the speed at which they are moving towards, or away from, the Earth. **(6 marks)**

Score / 12

How well did you do?

| 0–6 Try again | 7–13 Getting there | 14–20 Good work | 21–26 Excellent! |

For more information on this topic, see pages 50–51 of your Success Revision Guide.

Expanding Universe and Big Bang

Multiple-choice questions

Choose just one answer: A, B, C or D.

1 What observation supported the hypothesis that the Universe was initially very small and very hot? **(1 mark)**
- A The Big Bang
- B CMBR
- C red-shift
- D blue-shift

2 What attractive force prevents the Solar System from expanding? **(1 mark)**
- A blue-shift
- B The Big Bang
- C cosmic background
- D gravity

3 When do scientists think that the Big Bang happened? **(1 mark)**
- A 14 000 years ago
- B 14 000 000 years ago
- C 14 000 000 000 years ago
- D 14 000 000 000 000 years ago

4 What is a hypothesis? **(1 mark)**
- A a proven explanation for a phenomena
- B evidence to support a theory
- C a personal opinion
- D an explanation that takes account of limited data

5 The observations that Hubble made led him to conclude that: **(1 mark)**
- A the further away a galaxy is, the faster it is moving away
- B all galaxies are moving away at a constant speed
- C nearby galaxies are moving away faster than distant ones
- D most galaxies are stationary

Score / 5

Short-answer questions

1 Draw a line to match each of Hubble's observations with the appropriate conclusion. **(3 marks)**

Observations	Conclusions
The distances to spiral nebulae are greater than the size of the Milky Way.	The further away a galaxy is, the faster it is moving.
The light from all the distant galaxies is red-shifted.	All the distant galaxies are moving away from us.
The further away the galaxy is, the bigger the red-shift.	Spiral nebulae are outside the Milky Way and are likely to be distant galaxies.

2 Use the words provided to complete the passage below. You may need to use some words more than once. **(6 marks)**

> theory   radiation   confidence   hypothesis

The _____ that the Universe started with a massive explosion led to the following testable _____: The explosion produced _____ that would now be found in the microwave region of the spectrum. When data was found to support the _____, it increased scientists' _____ that the original _____ provided a reliable description of how the universe began.

Score / 9

GCSE-style questions

Answer all parts of all questions. Continue on a separate sheet of paper if necessary.

1 The following equation can be used to relate the red-shift to the speed (v) of a galaxy, where c (the speed of light) is 3×10^8 m/s:

$$\frac{\text{change in wavelength}}{\text{wavelength}} = \text{red-shift} = \frac{v}{c}$$

a) A galaxy gave a red-shift of 0.05

Calculate its speed. **(2 marks)**

b) If the red-shift in part **a)** was observed with radio waves of wavelength 5 m, calculate the change in wavelength observed. **(2 marks)**

c) The galaxy in part **a)** was 500 million light years away from Earth.

i) Tick the formula that shows the correct way to calculate this distance in metres. **(1 mark)**

- **A** $500 \times 10^6 \times$ speed of light $\times$ seconds in a year ☐
- **B** $500 \times 10^6 \times$ speed of light $\times$ days in a year ☐
- **C** $500 \times$ speed of light $\times$ seconds in a year ☐

ii) Use the calculation you have selected to show that 500 million light years equates to approximately 5×10^{24} m. **(2 marks)**

d) Describe what is meant by the 'Universe' and how red-shift data can be used to estimate its age. **(3 marks)**

Score   / 10

How well did you do?

| 0–6 | Try again | 7–12 | Getting there | 13–18 | Good work | 19–24 | Excellent! |

For more information on this topic, see pages 52–53 of your Success Revision Guide.

Distance, Speed and Velocity

Multiple-choice questions

Choose just one answer: A, B, C or D.

1. The distance moved per unit of time is the: **(1 mark)**
   A velocity
   B speed
   C magnitude
   D displacement

2. What measure is given by the average speed multiplied by the time taken? **(1 mark)**
   A velocity
   B acceleration
   C distance
   D vector

3. What is meant by velocity? **(1 mark)**
   A the speed travelled
   B the speed over a certain distance
   C the speed in a given direction
   D the distance travelled in a certain time

4. Vectors have: **(1 mark)**
   A magnitude and direction
   B direction and speed
   C magnitude and speed
   D magnitude and distance

5. Which of the following formulae is correct? **(1 mark)**
   A velocity = $\dfrac{\text{speed}}{\text{time}}$
   B speed = $\dfrac{\text{displacement}}{\text{time}}$
   C velocity = $\dfrac{\text{displacement}}{\text{time}}$
   D velocity = displacement × time

Score / 5

Short-answer questions

1. Calculate the speed of a man who runs 200 m in 40 s. **(1 mark)**

   ....

2. Graphs **A–D** show how your displacement from home (*y*-axis) varies with time (*x*-axis). Match each graph with the appropriate description from the list below. **(4 marks)**

   A   B   C   D

   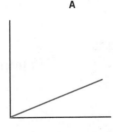 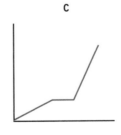 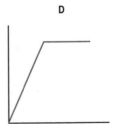

   1. You leave home by car to visit a friend. ....

   2. You walk to the bus-stop and wait for a bus, which takes you on your journey. ....

   3. You move away from home on foot. ....

   4. You return home by car. ....

Score / 5

GCSE-style questions

1 A car moving at 20 m/s decelerates to rest in 5 s after the brakes act, when a pedestrian steps into the road. The driver first takes 1 second to react before the brakes act.

a) On the grid alongside draw a velocity–time graph to represent this motion. **(4 marks)**

b) i) Label the part of the graph that represents the 'thinking distance'. **(1 mark)**

ii) State the thinking distance in metres. **(1 mark)**

c) i) Label the part of the graph that represents the 'braking distance'. **(1 mark)**

ii) Calculate the total braking distance. **(1 mark)**

d) Calculate the rate of deceleration whilst braking (assume deceleration was constant). **(2 marks)**

e) Describe how the graph would differ if the braking power was greater. **(1 mark)**

2 a) A car emerging from a side road at 4.5 m/s increases its speed to 17.2 m/s 2.5 s later. Calculate the acceleration of the car. Give your answer to 2 significant figures. **(3 marks)**

b) Describe what feature of a velocity–time graph would indicate that the car subsequently maintained its velocity. **(1 mark)**

Score / 15

How well did you do?

0–8 Try again | 9–15 Getting there | 16–21 Good work | 22–29 Excellent!

For more information on this topic, see pages 58–59 of your Success Revision Guide.

Forces

Multiple-choice questions

Choose just one answer: A, B, C or D.

1. What can be said about the forces acting on an object falling at a constant velocity? **(1 mark)**
   - A there are no forces acting
   - B all the force of gravity is greater than air resistance
   - C the drag force is the resultant force
   - D there is no resultant force

2. The force acting on you due to the Earth is called your: **(1 mark)**
   - A mass
   - B weight
   - C reaction force
   - D gravity

3. Which force acts to resist the motion of one solid surface across another? **(1 mark)**
   - A thrust
   - B lift
   - C friction
   - D gravity

4. Which of these statements is true for a skydiver the moment her parachute opens? **(1 mark)**
   - A air resistance > weight
   - B weight = air resistance
   - C air resistance < weight
   - D air resistance > gravity

5. Without a resultant force a moving object will: **(1 mark)**
   - A decelerate
   - B be stationary
   - C continue at a constant velocity
   - D lose mass

Score / 5

Short-answer questions

1. Complete the following statements about forces.

   a) The force on an object due to the Earth's gravitational pull is called its

   .... **(1 mark)**

   b) The units used to measure force are .... **(1 mark)**

   c) Newton's First Law of Motion states that... **(2 marks)**

2. a) On a separate piece of paper, draw a diagram of an aircraft in level flight at constant speed and label the four principal forces acting on it. **(4 marks)**

   b) State which pairs of forces must be equal to each other. **(2 marks)**

   c) State what an aeroplane must weigh if it is kept in level flight by a lift force of 10 kN. **(1 mark)**

Score / 11

GCSE-style questions

Answer all parts of all questions. Continue on a separate sheet of paper if necessary.

1 The Earth's surface gravitational field strength (g) can be taken as 10 N/kg.

a) State what a man of mass 70 kg must weigh on the Earth's surface. (1 mark)

b) On the Moon the gravitational field strength is 1.7 N/kg.

i) What would the same man's mass be on the Moon? (1 mark)

ii) What would the same man's weight be on the Moon? (1 mark)

2 The following data were obtained in an experiment where an increasing force was applied to a suspended spring.

Force applied (N)	Length of spring (cm)
0	15.0
1	17.1
2	19.1
3	20.9
4	23.0
5	24.9
6	27.1
7	29.2
8	32.0
9	36.0

a) On the graph paper above, plot a graph of force applied against length of spring. (4 marks)

b) Use your graph to give the length of the spring when a force of 3.5 N is applied. (1 mark)

c) When an object of unknown mass is hung from the spring it extends by 11.5 cm. Calculate the mass of the object. (3 marks)

Score / 11

How well did you do?

| 0–8 | Try again | 9–14 | Getting there | 15–20 | Good work | 21–27 | Excellent! |

For more information on this topic, see pages 60–61 of your Success Revision Guide.

Acceleration and Momentum

Multiple-choice questions

Choose just one answer: A, B, C or D.

1. Which one of the following is NOT a vector quantity? **(1 mark)**
   - A force
   - B speed
   - C momentum
   - D acceleration

2. The purpose of an airbag or crumple zone in a car is to increase the time taken for the driver to stop. What is the advantage of this? **(1 mark)**
   - A it reduces the driver's momentum
   - B it reduces the driver's energy
   - C it reduces the force on the driver
   - D it reduces the energy of the impact

3. Calculate the momentum of a 60 kg person running at 5 m/s. **(1 mark)**
   - A 12 kgm/s
   - B 300 kgm/s
   - C 120 kgm/s
   - D 300 kgm/s^2

4. What will a resultant force always cause? **(1 mark)**
   - A a change in direction
   - B a change in energy
   - C a change in speed
   - D a change in velocity

5. A tennis player returns a serve. How does the force on the ball relate to its momentum? **(1 mark)**
   - A the force is the rate of change of momentum of the ball
   - B the force is the overall change of momentum multiplied by the time taken to cause the change
   - C the force is the momentum of the ball as it is hit
   - D the force is the momentum of the ball leaving the racket divided by the time taken to cause the change

Score / 5

Short-answer questions

1. A shopping trolley, which is initially stationary on a smooth, horizontal surface, is given a steady push.

   a) Ignoring the effects of friction in the trolley wheels, describe what happens to the trolley. **(2 marks)**

   b) Describe the effect of the following changes.

   i) Increasing the mass of the trolley. **(1 mark)**

   ii) Increasing the strength of the push. **(1 mark)**

2. Newton's Second Law refers to the effect of an unbalanced force on an object. Ring the two phrases that could be used to correctly complete the sentence:

   The magnitude of the resultant force on an object is equal to the object's...

   final velocity / mass x acceleration / momentum x acceleration / rate of change of momentum / mass x change in velocity

   (2 marks)

Score / 6

Answers

Abbreviations used
; separates marking points
ORA or reverse argument
OWTTE or words to that effect

For questions marked ✐, where marks are awarded for the quality of written communication, model answers have been provided. The model answers would score the full 6 marks available. If you have made most of the points given in the model answer and communicated your ideas clearly, in a logical sequence with few errors in spelling, punctuation and grammar, you would get 6 marks. You will lose marks if some of the points are missing, if the answer lacks clarity and if there are serious errors in spelling, punctuation and grammar.

Pages 4–5 The Electricity Supply

Multiple-choice questions
1. B 2. C 3. D 4. A 5. C

Short-answer questions
1. a)

   b) i) electrical ii) light iii) heat
2. $\frac{8.4}{24} \times 100 = 35\%$

   (1 mark for calculation; 1 mark for answer)

GCSE-style questions
1. a) Passive solar cells use the heat from the sun to directly heat water; active solar cells transfer solar energy to electrical energy.
   b) **Advantages may include:** domestic solar panels can save on electricity bills; 'clean' energy. **Disadvantages may include:** expensive to install; UK weather unreliable.
   c) Efficiency is a measure of the amount of useful energy (electricity) produced from the total energy input.
   d) $\frac{310}{1000} \times 100 = 31\%$
   e) i) Energy is not destroyed, it is converted / transferred into different forms.
      ii) During the transfer process, some energy is wasted in 'non-useful' forms; such as heat energy and kinetic energy.
   f) **Answers may include:** increasing efficiency of energy transfer (e.g. in solar cells); reducing construction / installation costs; government initiatives to encourage use of renewable energy over fossil fuels.

Pages 6–7 Generating Electricity

Multiple-choice questions
1. A 2. D 3. C 4. B 5. D

Short-answer questions
1. a) **A** Furnace; **B** Turbine; **C** Generator; **D** Transformer; **E** Pylon / National Grid.
   b) i) Turbines allow water, steam or air to drive around the blades; turning the kinetic energy of the moving water, steam or air into rotational kinetic energy of the turbine to turn the generator.
      ii) Generators convert kinetic energy to electrical energy.
   c) They have formed over millennia from the decayed remains of pre-historic plants and animals.

GCSE-style questions
1. a) i) Biofuels produce carbon dioxide emissions when they are burned.
      ii) **Answers may include:** because biofuels are carbon neutral; they produce carbon dioxide emissions when burned but these are balanced / offset by growing plants to produce the biofuels; because the plants photosynthesise and remove carbon dioxide from the atmosphere.
      iii) **Any two from:** wood; straw; manure; sugar.
   b) **Answers may include:** availability of fuel; availability of water; impact on local environment; local infrastructure (roads, etc.); access by emergency services.
   c) voltage; rotates; changing.
   d) **Answers may include:** safety of plant / workers; radiation levels in local area; disposal of waste; how easily it could be decommissioned.

Pages 8–9 Renewable Sources of Energy

Multiple-choice questions
1. B 2. A 3. C 4. C 5. A

Short-answer questions
1. Renewable energy resources will not run out (will last forever, or at least for millennia); non-renewable energy sources are finite and will eventually run out (they cannot be replaced within a lifetime).
2. **Renewables – any three from:** hydroelectric power, biomass, tidal, wave, solar, geothermal, wind. **Non-renewables – any three from:** oil, gas, coal, nuclear
3. a) **Answers might include:** water reserves can diminish during droughts; land (habitats) have to be flooded to create reservoirs.
   b) **Answers might include:** tidal dams can destroy marine habitats; dams can stop fish migrating.
   c) **Answers might include:** power cannot be produced at night; poor weather conditions can limit power production.
   d) **Answers might include:** power can only be produced if there is sufficient wind to move the turbines; turbines can be frightening for livestock; some people consider turbines too noisy / ugly.
   e) **Answers might include:** volcanic action could damage the power station.

GCSE-style questions
1. a) They can give a lot of output power by allowing the water to run downhill; which drives turbines and generators placed there; the ease with which the water flow can be controlled enables a rapid response to demand.
   b) Stored GPE is converted to KE; KE is converted to electrical energy.
2. a) **Any two from:** effect on fish; effect of higher water levels on environment; spoils view.
   b) **Any two from:** cheaper electricity; job creation; 'clean' electricity.
3. ✐ Energy reaches the Earth from the Sun in the form of light and heat energy. The light from the Sun is captured by green plants and converted into useful energy by photosynthesis. This energy is then passed down the food chain when consumers eat the plants and eat each other. As fossil fuels, like coal and oil, are formed from the decayed remains of plants and animals, the energy in these fuels originally came from the Sun. For many renewable energy resources, the Sun can also be seen as the original source of energy. Solar panels capture the Sun's energy directly. However, other renewable energy resources are dependent on the weather which is affected by the Sun. For example, wind turbines are reliant on wind, which is produced by convection currents caused by the Sun, and hydroelectrical power is dependent on rainfall, which is caused by the Sun evaporating water as part of the water cycle.

Pages 10–11 Electrical Energy and Power

Multiple-choice questions
1. B 2. A 3. C 4. B 5. A

Short-answer questions
1. $1000 \times 60\ s \times 60\ min = 3\ 600\ 000$ J **(1 mark for calculation; 1 mark for answer)**
2. $100 \times 60 \times 5 = 30\ 000$ J **(1 mark for calculation; 1 mark for answer)**
3. a) amps (A) b) Volts (V) c) Watts (W)
4. **Answers may include:** Don't leave lights on; use energy-saving light bulbs; don't leave computers, etc. on standby (Accept any other sensible answers.)

GCSE-style questions
1. a) 0.1 kW $\times 12$ hrs $\times 7$ days $\times 10$p $= 84$p **(1 mark for calculation; 1 mark for answer)**
   b) i) 100 W bulb (10% useful energy): $100\ W \times \frac{10}{100\%} = 10$ W;
      18 W bulb (56% efficiency): $18\ W \times \frac{56}{100\%} = 10.08$ W;

we get the same useful light output from the 18 W bulb but with less electrical power input.

   ii) 0.018 × 12 × 7 × 10 = 15p **(1 mark for calculation; 1 mark for answer)**
   iii) 84p − 15p = 69p
 c) ✏ In terms of sustainability, the decision to ban 100 W bulbs is the right one. More efficient bulbs require less electricity, which means less fuel consumption. Since fossil fuels account for the majority of electricity production in the EU, this will help to conserve limited fossil fuel stocks. Emissions of pollutant gases and carbon dioxide, produced by burning fossil fuels, will also be reduced. However, to make a proper comparison, you would need to know how energy efficient the manufacturing process of both bulbs is. If a lot of energy is required to make the new bulb, it might outweigh the benefits.

Pages 12–13 Electricity Matters

Multiple-choice questions
1. C   2. D   3. D   4. B   5. C

Short-answer questions
1. a) electrical; generated; alternates; cables / power lines; step-down
   b) **Advantages may include:** power stations can be built close to fuel reserves; pollution can be kept away from densely populated areas; there is flexibility of power distribution; if a power station fails, coverage to all areas can still continue. **Disadvantages may include:** expensive to provide and maintain overhead / underground cabling; unsightly pylons and power cables; power is wasted during distribution

GCSE-style questions
1. a) The current keeps changing direction.
   b) i) Without the transformers power losses in the transmission lines would be too great; stepping up to high voltage for transmission reduces the current in the lines; and hence the power loss as heat.
      ii) Transformers work using electromagnetic induction; this requires a changing current; a direct current does not change.
2. a) 100 000 × 50 = 5 000 000 W (= 5 MW) **(1 mark for calculation; 1 mark for answer)**
   b) 5 000 000 W × 0.9 = 4 500 000 W (= 4.5 MW) **(1 mark for calculation; 1 mark for answer)**
   c) $\frac{4.5 \text{ MW}}{230 \text{ V}} = 19.6 \text{ kA}$ ($I = \frac{P}{V}$) **(1 mark for calculation; 1 mark for answer)**
3. ✏ I agree with the German government because although nuclear fuel is cheap and clean to use there, in my opinion the safety issues by far outweigh these advantages. Nuclear fuel is cost effective because the fuel is not expensive and a small amount produces a large amount of energy. The energy transfer process does not produce carbon dioxide emissions like fossil fuels, so it does not contribute to the greenhouse effect and global warming. However, the radioactive waste that is produced remains dangerous to living things for thousands of years, so it has to be stored very carefully. In my view, this is not sustainable, as future generations will have to manage growing amounts of nuclear waste. There are also concerns about the safety of nuclear power stations in the events of a terrorist attack or a natural disaster like an earthquake. I do not think that it is worth taking these risks to save money. (Accept answers that disagree with decision, as long as supporting arguments are given.)

Pages 14–15 Particles and Heat Transfer

Multiple-choice questions
1. D   2. A   3. B   4. C   5. B

Short-answer questions
1. a) 0.2 × 4200 × 80 = 67 200 J **(1 mark for calculation; 1 mark for answer)**
   b) 25 × 4200 × 10 = 1 050 000 J **(1 mark for calculation; 1 mark for answer)**
   c) Water has a large specific heat capacity; which means it can transfer a large amount of heat even for a relatively small temperature change.

GCSE-style questions
1. a) 100°C (the water was boiling)
   b) To ensure a fair test; so that the only variable affecting the results of the experiment is the colour of the box.
   c) i) matt black   ii) silver
   d) The major mechanism of heat transfer is radiation; conduction to the surrounding air and convection currents in the air are the same for all boxes; black surfaces are the most efficient radiators; whereas reflective surfaces are the least efficient.
   e) Ensure the results are repeatable.

Pages 16–17 Describing Waves

Multiple-choice questions
1. B   2. A   3. D   4. A   5. B

Short-answer questions
1. wavelength = $\frac{\text{wave speed}}{\text{frequency}} = \frac{20}{2} = 10 \text{ cm}$ **(1 mark for calculation; 1 mark for answer)**
2. speed = 0.34 × 1000 = 340 m/s; distance = 340 × 5 = 1700 m **(1 mark for calculation; 1 mark for answer)**
3. A false   B true   C false

GCSE-style questions
1. a) vibrations / oscillations; energy; longitudinal; parallel
   b)

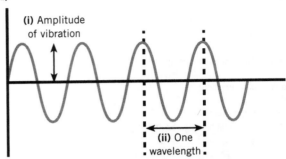

   (Accept any other accurate indications of wavelength, e.g. trough to trough)
   c) i) The amplitude would be less.
      ii) The frequency would be higher / waves would be closer together.
   d) Sound is a longitudinal wave; the diagram illustrates a transverse wave; the air particle displacement in a sound wave is parallel to the x-axis (horizontal axis).
   e) compressions; rarefactions; further apart; amplitude

Pages 18–19 Wave Behaviour

Multiple-choice questions
1. B   2. C   3. A   4. D   5. B

Short-answer questions
1. a) The large angle of reflection should be circled.
   b) A new line should be drawn making the angle of incidence and angle of reflection equal.
   c) When a wave is reflected, the angle of reflection is equal to the angle of incidence.

GCSE-style questions
1. a) At the boundary between the air and the water of the pool, the light is refracted away from the normal; so the bottom of the pool appears closer to the surface than it really is. **(1 mark for suitable diagram)**
   b) Light under the water's surface does not have to pass from one material to another in order to reach the observer's eyes; therefore the light is not refracted and doesn't change the observer's perception of the bottom of the pool.
2. When waves pass through a gap that is a similar size to the wavelength they diffract (spread out); Light has an extremely small wavelength compared to buildings, so cannot diffract around corners; Sound has a longer wavelength and can diffract, as well as reflect, around buildings.
3. a) wavelength = $\frac{\text{wave speed}}{\text{frequency}} = \frac{1500}{30\,000} = 0.05 \text{ m} = 5 \text{ cm}$ **(1 mark for calculation; 1 mark for answer)**
   b) 1500 × 0.5 = 750 m is the total distance travelled by the sound; therefore the distance to the submarine is 375 m.

Pages 20–21 Seismic Waves and the Earth

Multiple-choice questions
1. C   2. A   3. B   4. C   5. D

Short-answer questions
1. seismic; P; longitudinal; liquid; S; liquid; seismometers
2. convection
3. **Any two from:** jigsaw like match of shapes of continents; similar fossils on these continents, which are now thousands of miles apart; the magnetic alignment change in rock patterns many miles apart which reflect periodic variations in the Earth's magnetic field.

GCSE-style questions
1. a) i) and ii)

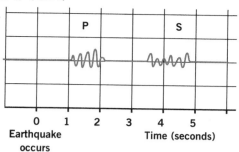

   b) $(3.5 - 1) \times 8 = 20$ km
   c) i) On the opposite side of the Earth.
      ii) B
   d) Scientists know that the Earth has a liquid core, because S-waves cannot travel directly though the Earth; By simulating earthquakes using controlled explosions they can trigger P-waves and S-waves; Monitoring these waves and analysing the data provides information about rock structure.

Pages 22–23 The Electromagnetic Spectrum

Multiple-choice questions
1. B   2. D   3. C   4. D   5. D

Short-answer questions
1. a) continuous; transverse; 300 000 000
   b) radio waves; microwaves; infrared; visible light; ultraviolet; X-rays; gamma rays.
2. The diagrams suggest that gamma rays have shorter wavelengths than X-rays; in fact, they overlap significantly in wavelength range.

GCSE-style questions
1. a) $3 \times \dfrac{10^8}{0.02} = 1.5 \times 10^{10}$ Hz **(1 mark for calculation; 1 mark for answer)**
   b) $3 \times \dfrac{10^8}{6} \times 10^{15} = 5 \times 10^{-8}$ m (= 50 nm) **(1 mark for calculation; 1 mark for answer)**
   c) i) 0.6 m; to 1 km
      ii) 250 m
2. a) Intensity is the amount of energy per second arriving at a square meter of surface.
   b) Moving closer to the source; increasing the output of the source.
   c) Red light has the lowest frequency in the visible range; Therefore the photons have less energy and are insufficient to cause the film to react; It is not the total energy arriving per second that counts, but the energy per photon.

Pages 24–25 Light, Radio Waves and Microwaves

Multiple-choice questions
1. A   2. C   3. D   4. C   5. B

Short-answer questions
1. The electrons oscillate in response to the signal, forming an alternating electrical current.
2. a) To prevent the waves escaping and possibly harming people.
   b) Toast should be crisp, so it needs infrared radiation to heat and warm the surface; Microwaves would penetrate and cook from inside to a degree.
3. image; lens; retina.

GCSE-style questions
1. a) ✏ In a microwave oven, the food is irradiated with microwaves. The food is rotated on a turntable to ensure even irradiation. The microwaves pass through glass and plastic packaging. The frequency of the microwaves used ensures that they are easily absorbed by water molecules in food. The absorbed microwaves increase the kinetic energy of the water molecules, which heats the food. Conduction and convection spread the heat throughout the food.
   b) There is water content in the Earth's atmosphere; microwaves of the wavelength range that is absorbed by water are therefore not suitable for transmission because they will be absorbed.
   c) Moisture in the atmosphere will absorb microwaves; so by measuring the amount of microwaves absorbed, the amount of water in the atmosphere can be calculated; An atmosphere with a high water content is more humid and likely to lead to rain or stormy weather.

2. a) high energy infrared; low energy ultraviolet.
   b) High energy ultraviolet; is absorbed in the upper levels of the atmosphere by ozone / infrared; is absorbed by the atmosphere.

Pages 26–27 Wireless Communications 1

Multiple-choice questions
1. D   2. C   3. B   4. A   5. B

Short-answer questions
1. Regeneration – The process of cleaning up a digital signal; Amplification – Increasing the size of a signal; Noise – Unwanted signals; Modulation – The changing of a signal that codes the information being sent.
2. A false B true C true D true E false F true

GCSE-style questions
1. Electromagnetic waves are diffracted by objects with dimensions of a similar size to one wavelength, causing them to spread out; radio waves, with wavelengths of a few metres or more, are diffracted by buildings, and natural features of the landscape; so reception does not have to be on a line of sight with the transmitter; microwaves, with wavelengths of a few centimetres, are not so readily diffracted; and so transmitters and receivers have to be in 'line of sight'.
2. a) A 3 B 1 C 5 D 2 E 4
   b) A code made up of just two values: 0 and 1.
   c) i) unwanted signals; that distort the original signal.
      ii) ✏ The claim about unwanted noise is justified because digital signals give a better quality reception than analogue signals. This is because any noise picked up during transmission can be more easily removed. The original values can easily be restored because each pulse can only be a 0 or a 1. The process of cleaning up the signal is called regeneration. Analogue signals can be amplified, but any noise is amplified too.

Pages 28–29 Wireless Communications 2

Multiple-choice questions
1. B   2. A   3. B   4. A

Short-answer questions
1. a) **Any two from:** mobile phone; printer; Internet; local computer network (Accept any other sensible answer.)
   b) **Any three from:** ease of connectivity; portability of various devices; no wires needed; 24 hour access.
2. A3; B1; C2.

GCSE-style questions
1. a) Microwaves are used for heating / cooking food.
   b) The power involved for mobile phones is much less than that involved in a microwave oven; so the effects would be negligible.
   c) The effects of microwaves from mobile phones could be accumulative; or take a long time to become evident.
   d) **Any two from:** frequency and length of use of mobile phone; other occupational or lifestyle choices that might be relevant (e.g. smoking); difficulties in finding a matched group of people who are infrequent users of mobile phones (Accept any other sensible answer.)
   e) There is potential for bias if the scientists are being paid by a mobile phone company.
   f) Your answer should clearly state your opinion; assess the potential risk (the heating effect of microwaves); assess the vulnerability of those most at risk (young people); assess exposure (very low intensity, for prolonged periods of time); assess the heating effect at such levels; and assess the ethics of the proposal.

Pages 30–31 Infrared

Multiple-choice questions
1. A   2. C   3. C   4. B   5. A

Short-answer questions
1. If metal is heated to a high enough temperature it will emit light (red then white).
2. A false B false C true D true E true
3. glass; refractive index; total internal.

GCSE-style questions
1. a) Infrared
   b) i) Infrared has a higher frequency than radio waves and microwaves.
      ii) Infrared signals in fibre optic cables experience less interference; A stream of data can be transmitted very quickly; Digital signals can be multiplexed, so that lots of different signals can be sent along the fibre at the same time.

2. a) Nocturnal animals come out at night, so it would be difficult to capture them in the dark with a standard camera; The animal will be warmer than its surroundings so it will be visible in the infrared region at night.
   b) Infrared is outside the visible spectrum; so different visible colours (false colour) are used to depict different temperature ranges indicated by the infrared wavelength detected.
   c) A survivor trapped under rubble will be warmer than the surroundings; The warmth of their body will be visible in the infrared region; A thermal imager picks up on this temperature difference by observing the infrared region; **Any one limitation:** Use of the technique can be complicated in the presence of other sources of heat; It is also limited by the depth of material through which the infrared can be detected.

Pages 32–33 The Ionising Radiations

Multiple-choice questions
1. C   2. A   3. B   4. D

Short-answer questions
1. **Answers might include:** detecting forged bank notes; forensic detection of powders / fluids (Accept any other sensible answer.)
2. a) It knocks electrons out of the atoms electron shells, so the atom becomes charged (ionised).
   b) alpha; beta; gamma
   c) i) gamma ii) alpha
   d) alpha – a helium nucleus; beta – an electron.

GCSE-style questions
1. a) The risk to the radiographer would be too big as they would be being exposed regularly; leading to a large cumulative exposure
   b) A patient has a single exposure, which carries a small risk; The risk is offset by the benefit to the patient from the information obtained; allowing diagnosis / better treatment of their medical problem.
2. a) Ionising radiation causes atoms to become charged, so they are more likely to take part in chemical reactions; If these atoms are inside the body's cells, this could damage or kill the cells; If the DNA of a cell is damaged it may mutate; Mutations can cause cells to grow out of control and become cancerous.
   b) The scientist should select a wide range of volunteers; to gain a picture of a cross-section of society.
   c) People within each group should be matched according to age, sex, ethnicity and lifestyle; relevant exposure to other known carcinogens (e.g. smoking); frequency and duration of sun bed use.
   d) A control group would be a group of people who do not use sun beds; this provides a good comparison so that other factors can be ruled out.
   e) A correlation shows a pattern / relationship; but it is not proof of cause.

Pages 34–35 The Atmosphere

Multiple-choice questions
1. D   2. D   3. B   4. A   5. A

Short-answer questions
1. a) Low energy infrared
   b) Light and high energy infrared
2. **A** false **B** false **C** true **D** true **E** false **F** true

GCSE-style questions
1. a) **Any two from:** extreme weather conditions; rising sea levels; expansion of oceans / flooding of lowlands; some regions no longer able to grow food crops (Accept any other sensible answer.)
   b) Greenhouse gases behave like the glass in a greenhouse; They allow high energy infrared from the sun into the atmosphere, but prevent low energy infrared from escaping, i.e. they trap heat.
   c) **Any two from:** increased combustion of fossil fuels in cars / vehicles; increased burning of fossil fuels to provide electricity; deforestation (less carbon dioxide being removed from the atmosphere by photosynthesis) (Accept any other sensible answer.)
   d) **Any one from:** There have been other periods of extreme weather (i.e. ice ages and droughts) in the history of the Earth; it is difficult to specify what temperature increase / duration constitutes global warming; there are political issues involved (Accept any other sensible answer.)
2. a) The ozone layer is a layer of ozone gas in the upper atmosphere; It absorbs some UV, limiting the amount that reaches Earth.
   b) Greater UV exposure will lead to greater incidence of sunburn, skin cancer and cataracts.
   c) The concentration of CFCs in the atmosphere increased; this reacted with the ozone gases, damaging the ozone layer.

Pages 36–37 The Solar System

Multiple-choice questions
1. C   2. C   3. B   4. A   5. B

Short-answer questions
1. a) Inner planets are relatively small and rocky / solid; Outer planets are big and gaseous.
   b) the asteroid belt/asteroids.
2. Sun – The star at the centre of the Solar System; Moon – A natural satellite in orbit around a planet; Asteroid – A rock, up to about 1 km in diameter, in orbit around the Sun; Comet – Accumulated ice and dust, with a nucleus that vaporises when it comes close to the Sun; Meteor – A small rock, usually from a comet, entering the Earth's atmosphere.

GCSE-style questions
1. a)

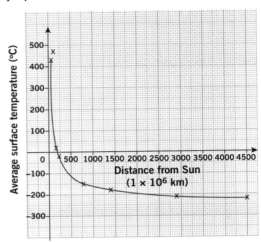

(1 mark for scale and label on *y*-axis; 1 mark for scale and label on *x*-axis; 1 mark for accurately plotted points; 1 mark for line of best fit.)

   b) The further a planet is from the Sun, the colder its average surface temperature; this is because radiation from the Sun is less intense at greater distances.
   c) Venus
   d) The atmosphere on Venus may have a greenhouse effect. (Accept any other sensible answer.)

Pages 38–39 Space Exploration

Multiple-choice questions
1. B   2. D   3. B   4. A   5. C

Short-answer questions
1. Earth; Mars; orbit; compacted.
2. X-ray – In orbit above the Earth's atmosphere; Radio wave – Ground-based; Microwave – Ground-based; Infrared – In orbit above the Earth's atmosphere.

GCSE-style questions
1. a) i) Because of the huge distances involved; it would take too long for the radio signals to travel to the receiver.
      ii) **Answers may include:** you cannot make immediate changes; there is always a delay between data received and responses.
   b) **Answers may include:** manned missions provide more scope for responding to unexpected circumstances; manned missions inspire public interest.
   c) **Answers should include any four from:** safety; provision of life support; regulation of temperature; shielding from cosmic rays; maintaining astronaut fitness; expense; time involved (Accept any other sensible answers.)
2. Any two advantages and any two disadvantages. **Answers may include:** Orbiting telescopes do not have to suffer the effect of the Earth's atmosphere on the light, so are more sensitive and their images are clearer; orbiting telescopes suffer less from Earth-based interference (such as light pollution); orbiting telescopes are more expensive; orbiting telescopes are more difficult to maintain (Accept any other sensible answers.)

Pages 40–41 A Sense of Scale

Multiple-choice questions
1. B 2. A 3. C 4. B 5. D

Short-answer questions
1. **A** false **B** true **C** true **D** false **E** true
2. B; A; D; E; C

GCSE-style questions
1. a) i) Parallax is the change in angle of viewpoint of an object as the observer moves position; The angle depends on the object's distance away (smaller angle for greater distance).
   ii) Observation has to be made from two points as far apart as possible; this means at two points six months apart in the Earth's orbit.
   b) **Any two from:** size; power output (or temperature); distance away.
   c) Just as the colour of the rod changes with temperature; the colour of a star indicates its surface temperature; The closer the colour is to the violet end of the visible spectrum the hotter the star.
   d) From their brightness and / or red-shift.
   e) For large distances light takes a significant period of time to reach Earth; the light the astronomer observes left the star some time earlier.

Pages 42–43 Stars

Multiple-choice questions
1. A 2. D 3. B 4. C 5. D

Short-answer questions
1. a) gravitational force
   b) The radiation and kinetic energy of the material inside the star exerts an outward pressure; this balances the gravitational attraction.
2. **A** gas cloud **B** main sequence **C** red supergiant **D** supernova **E** neutron star or black hole

GCSE-style questions
1. a) A star starts to become unstable when the hydrogen nuclei that 'fuel' fusion start to run out. In a small star, like our Sun, the star cools, becoming redder, and expands to form a red giant. The core contracts and the outer are layers are lost (forming the planetary nebula). The remaining core becomes a hot and dense white dwarf, which will eventually cool to become a black dwarf.
   b) As a very massive star swells it forms a red supergiant; that eventually explodes (a supernova); the collapse of the red supergiant core can lead to a neutron star or even a black hole after the supernova explosion.
2. The core of a star must be many millions of degrees so that the hydrogen nuclei have enough kinetic energy; to overcome their repulsion and fuse to form helium nuclei; releasing energy as electromagnetic radiation.
3. 1604 − 20 000 = −18 396, explosion occurred in 18 396 BC

Pages 44–45 Galaxies and Red-Shift

Multiple-choice questions
1. B 2. C 3. B 4. D 5. C

Short-answer questions
1. a) The further away the galaxy is, the faster it moves away.
   b) The measured red-shift can be used to determine distance
2. expanding; Big Bang; further; greater; faster; away; Doppler effect; blue.

GCSE-style questions
1. a) **Answers may include:** Insufficient data may be available to distinguish between alternatives; personal opinion may influence a scientist's choice between options (Accept any other sensible answer.)
   b) **Any three from:** presentation at scientific conferences; publication in scientific journals; peer review; reproduction of results by other scientists.
   c) A; B; E.
2. Light from distant stars can be analysed with a spectrometer. Each element produces a unique set of spectral lines (absorption spectrum), so the lines can be used to identify which elements are present. Dark lines represent wavelengths that have been absorbed by elements in the outer regions of the star. Bright lines are wavelengths emitted by atoms of hot gases. The elements present in the outer regions of the star will each absorb specific wavelengths, so the absorption spectrum produced. If the star is moving away from the Earth the absorption spectrum is shifted towards longer wavelengths (red shift). The extent of the shift is indicative of the speed.

Pages 46–47 Expanding Universe and Big Bang

Multiple-choice questions
1. B 2. D 3. C 4. D 5. A

Short-answer questions
1. The distances to spiral nebulae are greater than the size of the Milky Way. – Spiral nebulae are outside the Milky Way and are likely to be distant galaxies; The light from all the distant galaxies is red-shifted. – All the distant galaxies are moving away from us; The further away the galaxy is, the bigger the red-shift – The further away a galaxy is, the faster it is moving.
2. theory; hypothesis; radiation; hypothesis; confidence; theory.

GCSE-style questions
1. a) $0.05 \times (3 \times 10^8) = 1.5 \times 10^7$ m/s (**1 mark for calculation; 1 mark for answer**)
   b) $0.05 \times 5 = 0.25 = 25$ cm (**1 mark for calculation; 1 mark for answer**)
   c) i) A
      ii) $500 \times 10^6 \times (3 \times 10^8) \times 365$ days $\times 24$ hours $\times 60$ minutes $\times 60$ seconds $= 4.7 \times 10^{24}$ m (**1 mark for calculation; 1 mark for answer**)
   d) The Universe is everything that exists (there is nothing outside the Universe); The age of the Universe can be estimated by using red-shift data to look at the distance of objects from Earth and the speed at which they are moving away from Earth; this information can be used to estimate the time since all distant objects first separated and started moving apart (i.e. when the Big Bang took place).

Pages 48–49 Distance, Speed and Velocity

Multiple-choice questions
1. B 2. C 3. C 4. A 5. C

Short-answer questions
1. 5 m/s
2. **A** 3; **B** 4; **C** 2; **D** 1

GCSE-style questions
1. $330 \times 4 = 1320$ m (**1 mark for calculation; 1 mark for answer**)
2. a) $\frac{400}{50} = 8$ m/s (**1 mark for calculation; 1 mark for answer**)
   b) 0 m/s; velocity is a vector and the displacement after a whole lap is zero.
3. a) 3 m/s $\times$ 60 s = 180 m (**1 mark for calculation; 1 mark for answer**)
   b)

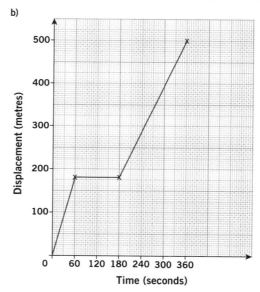

(**1 mark for scale and label on *y*-axis; 1 mark for scale and label on *x*-axis; 1 mark for accurately plotted points; 1 mark for line of graph.**)

   c) $\frac{500 \text{ m} - 180 \text{ m}}{180 \text{ s}} = \frac{320 \text{ m}}{180 \text{ s}} = 1.8$ m/s (**1 mark for calculation; 1 mark for answer**)
   d) $\frac{500 \text{ m}}{(6 \times 60 \text{ s})} = \frac{500 \text{ m}}{360 \text{ s}} = 1.4$ m/s (**1 mark for calculation; 1 mark for answer**)

Pages 50–51 Speed, Velocity and Acceleration

Multiple-choice questions
1. B 2. B 3. B 4. A 5. B

Short-answer questions
1. **A** travelling at a constant velocity **B** decelerating **C** travelling at a lower constant velocity **D** accelerating **E** stationary

2. Indication of a stationary object – Horizontal line on distance–time graph; Acceleration – Gradient of velocity–time graph; Distance travelled – area under velocity–time graph; Velocity – gradient of distance–time graph.

GCSE-style questions

1. a)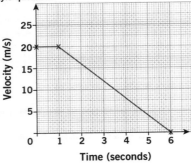

    (1 mark for scale and label on *y*-axis; 1 mark for scale and label on *x*-axis; 1 mark for accurately plotted points; 1 mark for line of graph.)

    b) i) area under the graph for the first second shaded and labelled thinking distance
        ii) 20 m
    c) i) area under the graph between 1 and 6 seconds shaded and labelled braking distance.
        ii) $\frac{20 \times 5}{2}$ = 50 m
    d) $\frac{20 m/s}{5 s}$ = 4 m/s²
    e) The line would have a steeper slope (greater gradient).

2. a) Change of speed = 17.2 – 4.5 = 12.7 m/s; acceleration = $\frac{12.7}{2.5}$ = 5.08 m/s² = 5.1 m/s² (2 s.f.)
    b) A horizontal straight line.

Pages 52–53 Forces

Multiple-choice questions
1. D   2. B   3. C   4. A   5. C

Short-answer questions
1. a) weight
    b) Newtons
    c) ...all objects remain at rest or move at constant velocity (speed in a straight line); unless acted upon by a resultant external force.
2. a) Diagram of aircraft with arrows showing vertical forces – weight down; lift upwards; horizontal forces – thrust forwards; drag (air resistance) backwards.
    b) lift = weight; thrust = drag
    c) 10 000 N (10 kN)

GCSE-style questions
1. a) 700 N   b) i) mass is 70 kg   ii) weight is 119 N
2. a)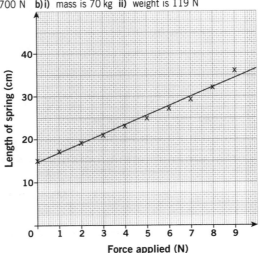

    (1 mark for scale and label on *y*-axis; 1 mark for scale and label on *x*-axis; 1 mark for accurately plotted points; 1 mark for line of graph.)

    b) 22.0 cm (+ or – 0.1 cm)
    c) total length of spring = 15 + 11.5 = 26.5; using graph force (weight) = 5.7N (+ or – 0.1N); mass = $\frac{weight}{g}$ = 0.57 kg

Pages 54–55 Acceleration and Momentum

Multiple-choice questions
1. B   2. C   3. B   4. D   5. A

Short-answer questions
1. a) The trolley starts to move, accelerating at a constant rate.
    b) i) The trolley would accelerate more gently.
       ii) The trolley would accelerate more quickly.
2. mass × acceleration; rate of change of momentum.

GCSE-style questions
1. a) Velocity = acceleration × time; 7 m/s **(1 mark for calculation; 1 mark for answer)**
    b) 280 kgm/s
    c) f = $\frac{280}{0.2}$ = 1400 N **(1 mark for calculation; 1 mark for answer)**
    d) Increasing the time reduces the force exerted on his body; there is therefore less likelihood of injury
2. a) Driver's initial momentum = 2100 kgm/s; 21 000 N
    b) Airbag / seatbelt
    c) Data showing reduced incidence of fatal injuries to drivers in road accidents; whether a seatbelt (or airbag) was used / speed of vehicle.

Pages 56–57 Pairs of Forces: Action and Reaction

Multiple-choice questions
1. B   2. B   3. B   4. C   5. C

Short-answer questions
1. velocity; the same; conserved; force; short; longer; size; opposite; action; reaction.

GCSE-style questions
1. a) Arrow from centre of Earth towards centre of Moon; arrow from centre of Moon towards centre of Earth; arrows of equivalent length
    b) The forces have the same magnitude; but act in opposite directions.
    c) Arrow from centre of person towards centre of Earth; arrow from centre of Earth towards centre of person; arrows of equivalent length.
2. a) momentum of car = 1000 × speed; momentum of truck = 2500 × speed in opposite direction; total momentum = 1500 × speed in direction of truck.
    b) total momentum = 22 500 kgm/s; total mass = 3500 kg; velocity = 6.4 m/s; in original direction of truck.

Pages 58–59 Work and Energy

Multiple-choice questions
1. C   2. C   3. A   4. A   5. B

Short-answer questions
1. a) Mass × gravitational field strength × height = 225 000 J **(1 mark for calculation; 1 mark for answer)**
    b)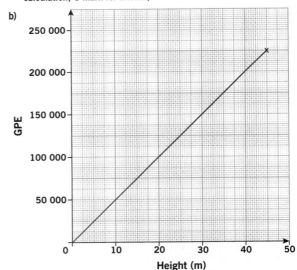

    (1 mark for scale and label on *y*-axis; 1 mark for scale and label on *x*-axis; 1 mark for line of graph.)

    c) i) Potential energy at half height = 112 500 J; Kinetic energy on impact = 112 500 J
       ii) $\frac{1}{2}$ × mass × velocity² = 112 500; 21.2 m/s

GCSE-style questions
1. a) ✎ At point 1, the rollercoaster has gravitational potential energy (GPE) but no kinetic energy (KE) because it is stationary. As it falls down the first drop, GPE transfers to KE so at point 2 the rollercoaster has less GPE than at the start, plus some KE. At point 3, the rollercoaster has the minimum GPE and maximum KE. As the roller coaster begins to climb the loop at point 4, some of this KE transfers to GPE.
   b) the starting height; energy.
   c) GPE change = mass × g × change in height; KE = GPE change; 875 kJ
   d) $\frac{1}{2}$ × mass × velocity^2 = 875 000; 26.5 m/s
   e) mass × g × change of height = ½ × mass × velocity^2; therefore velocity is independent of mass; so there would be no change to answer d).

Pages 60–61 Energy and Power

Multiple-choice questions
1. B   2. B   3. A   4. C   5. B

Short-answer questions
1. a) 100 N × 10 m = 1000 J
   b) $\frac{100 \text{ J}}{4 \text{ s}}$ = 250 W
2. a)

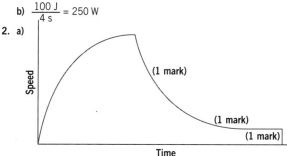

   b) balanced; terminal.

GCSE-style questions
1. a) Gain in GPE = 744 J; KE increases by 744 J in 1.2 s; 620 W (J/s)
   b) GPE increases by 744 J in 6 s; 124 W
2. a) 80 km/h = 22.2 m/s; KE = 740 kJ
   b) Work done = force × distance = 740 000; 7400 N
   c) i) thinking distance = speed × thinking time; 20 m
      ii) Braking distance reduced to 80 m; 9250 N
   d) i) driver's alertness
      ii) **Any one from:** road conditions; brake or tyre condition; vehicle load.
      iii) speed

Pages 62–63 Electrostatic Effects

Multiple-choice questions
1. C   2. D   3. B   4. B

Short-answer questions
1. Electrons; balloon; opposite; opposite; attraction.
2. The conductor must be fully insulated; otherwise the electrons readily flow to and from the conductor, preventing charge build up.

GCSE-style questions
1. a) ✎ In this experiment, the flour passes through the apparatus in the same way as it would in a flour mill. The flour slides down the chute and the friction against the surface of the chute causes a charge transfer. Charged flour builds up in the can, as indicated by the electroscope. The chute retains the opposite charge. In the same way, flour moving along chutes in a mill, results in an accumulating static charge separation.
   b) If the charge separation builds up sufficiently it will discharge with a spark; this spark could be sufficient to ignite a combustible material.
   c) A conductor can be connected between the chute and the can; this will safely discharge the static before it can build up sufficiently to pose a hazard.
   d) **Answers may include:** The refuelling of aircraft has the potential for charge separation in a combustible environment; connecting the plane and the tanker, or earthing both, prevents the charge from building up.

Pages 64–65 Uses of Electrostatics

Multiple-choice questions
1. A   2. D   3. A   4. C   5. B

Short-answer questions
1. negative; the same; repel; attracted; opposite; even; away.
2. The charge flows through the patient's chest between the defibrillator plates; The flow of charge causes the heart muscles to contract; This is used to treat patients with cardiac arrhythmia or whose hearts have stopped.

GCSE-style questions
1. a) **A** 3; **B** 1; **C** 5; **D** 2; **E** 4; **F** 6.
   b) The paper would need to be positively charged; the opposite charge will attract the toner to the paper.
2. Smoke particles in the exhaust stream pass through a charged grid; the particles become charged; the charged particles are attracted to the oppositely charged plates lining the walls of the chimney; the particles stick to the plates and clump together.

Pages 66–67 Electric Circuits

Multiple-choice questions
1. B   2. C   3. D   4. A

Short-answer questions
1. charge; continuous; current; electrons; positive; negative; amps; the same as.
2. a) 500 mA   b) 1.5 A   c) 0.6 A

GCSE-style questions
1. a) and b)

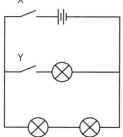

   c) If B can be switched off then there would be no current flowing through B; therefore, no current flowing through C as they are on the same piece of wire.
   d) The branch of the circuit with Bulbs B and C has twice the resistance of that with Bulb A; therefore, the current through B and C is 0.1 A; the total current is the sum of the currents through each branch; so it is 0.3 A.
   e)

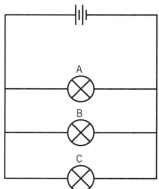

   f) Each bulb now has a current of 0.2 A; total current is therefore 0.8 A.

Pages 68–69 Voltage or Potential Difference

Multiple-choice questions
1. D   2. B   3. C   4. A   5. B

Short-answer questions
1. a) true   b) true   c) false   d) true   e) true   f) false
2. a) A_3 = 0.2 A   b) A_4 = 0.3 A   c) V_2 = 3 V   d) V_3 = 3 V

GCSE-style questions
1. a) One circuit diagram showing the two bulbs in series; correct symbols must be used; one circuit diagram showing the two bulbs in parallel; correct symbols must be used.

b) In the series circuit, both bulbs will be of equivalent brightness because they have the same current; in the parallel circuit, both bulbs will be of equivalent brightness because they have the same potential difference; the bulbs in series circuit will be less bright than those in the parallel circuit because the potential difference is divided between the bulbs.
c) i) **Answers may include:** something wrong with one of the components; user error.
ii) The student should repeat the experiment.
d) i) series ii) parallel
e) 1.5 V = 1.5 J/C; charge for 0.3 J = 0.2 C; 0.2 C/s = 0.2 A.

Pages 70–71 Resistance and Resistors

Multiple-choice questions
1. A   2. C   3. C   4. B   5. D

Short-answer questions
1. a) the free electrons
   b) The electrons are free to move if a potential difference is applied – free electrons move from negative to positive (actual current)
   c) In an insulator the electrons are tightly bound and not able to move if a potential difference is applied.
2. a) 0.3 A
   b) the current would fall to 0.15 A
   c) the current would increase to 0.6 A

GCSE-style questions
1. a) the lattice of ions
   b) electrons collide with the lattice ions and lose energy; the ions vibrate more, causing an increase in temperature
   c) resistance is the voltage per unit current; it is measured in ohms
2. a) 125 mA
   b) first bulb = 1 V; second bulb = 2 V
   c) i) first bulb = 375 mA; second bulb = 188 mA.
      ii) total current = 0.56 A; effective resistance = 5.3 Ω.

Pages 72–73 Special Resistors

Multiple-choice questions
1. C   2. C   3. A   4. C   5. C

Short-answer questions
1. a) As the filament temperature increases, the lattice ions vibrate more; this results in an increase in its resistance.
   b) In a thermistor, resistance decreases with increasing temperature.
   c)

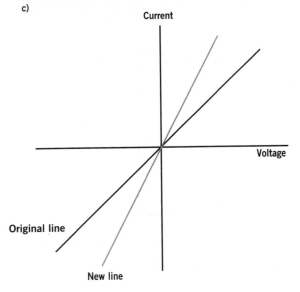

Original line

New line

GCSE-style questions
1. a) i) 10 Ω   ii) 12 Ω
   b) Non-Ohmic; because current is not directly proportional to applied voltage.

c)

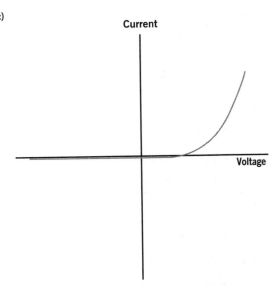

(1 mark for: current flow in forward direction for voltage above a threshold; 1 mark for: no current for reverse polarity; 1 mark for: steep gradient when current flowing in forward direction.)

d) To prevent current flow in one particular direction.
2. Resistance is dependent on the amount of light falling on the LDR; the greater the intensity, the less resistance.

Pages 74–75 The Mains Supply

Multiple-choice questions
1. C   2. B   3. B   4. A   5. A

Short-answer questions
1. a) 1 = earth wire; 2 = neutral wire; 3 = live wire; 4 = cable grip; 5 = fuse.
   b) Safety device that limits the current that can flow; current exceeding the rating causes heating sufficient to melt the thin wire within the fuse.
   c) 13 x 230 = 2990 W; 3.0 kW

GCSE-style questions
1. a) Earthing involves ensuring all exposed metal parts are connected to Earth; if a fault develops and the live wire touches any of these parts a large current flows safely to Earth; additionally, the fuse will melt ('blow') and cut the electricity supply.
   b) If there are no exposed metal parts then earthing is unnecessary as the user cannot come into contact with the live current; this is called double insulation.
2. a) Residual current is caused by a difference in the currents in the live and neutral wires, indicating a leakage to earth.
   b) The RCCB 'trips' (breaks the circuit) if this current is excessive; preventing the operator from receiving an electric shock.
   c) A circuit breaker is more sensitive than a fuse; and responds much more rapidly when a problem occurs.

Pages 76–77 Atomic Structure

Multiple-choice questions
1. C   2. A   3. D   4. D   5. B

Short-answer questions
1. a) electron: $\frac{1}{1840}$, –1; proton: 1, +1; neutron: 1, 0
   b) The number of electrons is equal to the number of protons.
2. a) 26 protons, 30 neutrons, 26 electrons
   b) 15 protons, 17 neutrons, 15 electrons
   c) 19 protons, 20 neutrons, 18 electrons

GCSE-style questions
1. a) The atom is made of positively charged material; it is embedded with negatively charged electrons; diagram should show a 'solid' looking sphere (pudding), studded with negative particles.
   b) C; E.
2. a) Isotopes of an element all have the same number of protons / atomic number; there are two different atomic numbers given (7 and 8).
   b) Isotopes are atoms of the same element with the same number of protons; and with varying numbers of neutrons (and therefore mass).
3. $^{218}_{84}$Po

Pages 78–79 Radioactive Decay

Multiple-choice questions
1. C 2. C 3. B 4. D 5. B

Short-answer questions
1. a) false b) true c) false d) true e) true f) false
2. a) 4000 Bq
   b) 4000 ÷ 2 ÷ 2 ÷ 2 = 500 Bq (**1 mark for calculation; 1 mark for answer**)

GCSE-style questions
1. a) Sam would get no change in the reading for both beta and gamma if a material like paper was used; the reading would partially decrease if a sheet of aluminium a few mm thick (or similar) was used because beta would stop; the reading would drop to almost zero if several cm of lead or a substantial thickness of concrete would be used as this is necessary to stop gamma readings.
   b) **Any two from:** handle materials with tongs; wear protective clothing; limit exposure time to the materials; (Accept any other sensible answer.)
   c) The time taken for half the radioactive atoms (or nuclei) to decay / for the activity to fall by half.
   d) 🖉 The half-life of an isotope is the average time taken for half the nuclei present to decay. As it an average, some atoms will decay faster and some will decay slower than the rate given as the half-life. Therefore, it is impossible to say when, exactly, an individual atom will decay. Both statements are true in this respect.
   e) False (external conditions do not affect the intrinsic decay rate)

Pages 80–81 Living with Radioactivity

Multiple-choice questions
1. B 2. C 3. D 4. D 5. B

Short-answer questions
1. a) Some of the elements present in our food have naturally occurring radioisotopes; Exposure of food to radiation does not make it radioactive.
   b) If our food is contaminated then we will take excess radioactive material into our bodies; Irradiation from within our bodies will result in much greater exposure / energy absorption and damage to our cells than exposure from outside.

GCSE-style questions
1. a) radiation that occurs in the environment.
   b) granite rock releases radon gas; radon can accumulate within closed spaces (such as buildings).
   c) 🖉 If radon gas is inhaled, it will then decay in the lungs. The decay process results in deposits of radioactive material. As a result, the lungs will be exposed to significant irradiation from within, leading to an increased risk of lung cancer.
2. a) 24 hours later is 13 half-lives; the activity will therefore drop by a factor of 2^{13} (i.e. 8192-fold).
   b) The background activity is approximately 2 Bq; this should be subtracted from the initial activity of 212 Bq.

Pages 82–83 Uses of Radioactive Materials

Multiple-choice questions
1. C 2. A 3. D 4. B 5. C

Short-answer questions
1. Smoke particles reduce the number of alpha particles reaching the detector so the alarm is triggered; Beta and gamma emissions are unsuitable as they would pass through the smoke.
2. Gamma; since the radiation must pass through a substantial thickness of soil.

GCSE-style questions
1. a) The level of activity reaching the detector is dependent on the thickness of the foil.
   b) Beta
   c) Steel would absorb too much beta radiation; so gamma would be more appropriate.
   d) Long half-life; so that the activity remains constant throughout the production process.
2. Exposure to radiation can potentially damage cells in the body or cause cancers to develop and so should be minimised for a healthy person; For a person who already has cancer the small risk of exposure is outweighed by the useful information gained to inform treatment planning.

Pages 84–85 Nuclear Fission and Fusion

Multiple-choice questions
1. C 2. D 3. C 4. B 5. B

Short-answer questions
1. a) i) fuel rods ii) moderator iii) reactor core iv) coolant v) turbines vi) neutrons
   b) Low level waste – Sealed into containers and put in landfill sites; Intermediate level waste – Mixed with concrete and stored in stainless steel containers; High level waste – Kept under water in cooling tanks.

GCSE-style questions
1. a) Repulsion
   b) i) high temperature; high pressure.
      ii) The fuel needed is hydrogen, which can be obtained from water; no radioactive waste is created.
2. a) $^{141}_{56}Ba + ^{92}_{36}Kr$
   b) 2 neutrons
   c) $^{235}_{92}U + ^{1}_{0}n \longrightarrow ^{140}_{54}Xe + ^{94}_{38}Sr + 2^{1}_{0}n$
   (**1 mark for each correct product**)

Pages 86–87 Refraction, Dispersal and TIR

Multiple-choice questions
1. C 2. C 3. A 4. C 5. A

Short-answer questions
1. a) The material through which the light travels.
   b) A change of speed / change of medium; the light must enter the medium at an angle (not along the normal).
2. a) i) and ii)

(**1 mark for refraction of red light at first boundary; 1 mark for refraction of red light at second boundary; 1 mark for positioning blue light below red; 1 mark for diverging blue light away from red.**)

   b) White light is made up of all visible frequencies; each of which refracts to a different extent, so you would see a spectrum of colours.

GCSE-style questions
1. a)

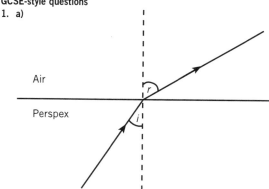

   b) i) $i = 35°$ ii) $r = 60°$ iii) $\frac{\sin 60}{\sin 35} = 1.7$ (+/− 0.1)
   c) i) A ii) 2×10^8 m/s
   d) $\sin(r) = 1.5 \times \sin 40$; $r = 75°$
   e) $\sin(c) = \frac{\sin 90}{1.5}$; $c = 42°$
   f) Total internal reflection.

Pages 88–89 Lenses

Multiple-choice questions
1. A 2. D 3. D 4. C 5. A

Short-answer questions
1. a)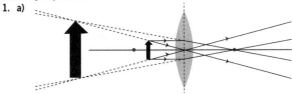

(1 mark for ray line through focal point; 1 mark for ray line through centre of lens; 1 mark for position of screen.)
   b) i) real ii) inverted

GCSE-style questions
1. a) The reciprocal of the focal length, i.e. $\frac{1}{\text{focal length}}$.
   b) Light from an object at a very great distance ('infinity'); is focussed onto a screen to measure the focal length (i.e. the distance from the lens to the screen); The power is then found by calculating $\frac{1}{\text{focal length}}$
   c) A; C.
2. a) dispersion; frequency.
   b) The blue light is focussed closest to the lens (i.e. in the shorter distance); because the refractive index is higher for blue light.

Pages 90–91 Seeing Images

Multiple-choice questions
1. B 2. C 3. C 4. A 5. C

Short-answer questions
1. a) the cornea; the lens. b) A

GCSE-style questions
1. a)

(1 mark for each correctly drawn ray line; 1 mark for correctly positioned image of arrow.)
   b) virtual; upright.
   c) approximately 3.3 times (accept 3.0 – 3.5)
2. The camera lens is moved in and out; to adjust the distance to the image/film/CCD.
3. ✏ When someone is short-sighted, it means that the light from a distant object is focussed in front of the retina. The power of the eye is too great / the minimum power of the eye is limited. A concave lens could be used to reduce the power, so that the focal length of the lens is increased to match the length of the eyeball / distance to the retina, bringing the distant object into focus.

Pages 92–93 Telescopes and Astronomy

Multiple-choice questions
1. C 2. A 3. C 4. B 5. C

Short-answer questions
1. a) The focal length of the eyepiece lens is shorter than that of the objective; therefore, its power is greater.
   b) $\frac{\frac{1}{0.75}}{\frac{1}{12}} = 16$
   c) focal length of objective = 1.33 m; focal length of eyepiece = 0.083 m; length of telescope = 1.41 m.

GCSE-style questions
1. a) Light is focussed at the focal point of the eyepiece using a curved mirror rather than a lens.
   b) It is easier to construct a large mirror than a large lens; there is no chromatic aberration with a mirror.
   c) A longer telescope allows for a weaker objective; and therefore a greater ratio between the power of the eyepiece and the objective; the magnification is increased.
2. a) Smaller
   b) One parsec is the distance to a star with a parallax angle of one second of arc.

Pages 94–95 Electromagnetic Effects

Multiple-choice questions
1. B 2. C 3. D 4. A

Short-answer questions
1. a) motor effect
   b) and c)

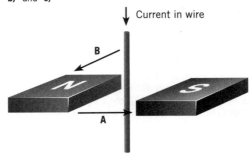

   d) The wire will be forced in the opposite direction

GCSE-style questions
1. a)

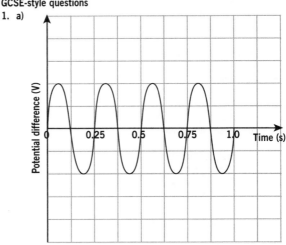

(1 mark for timescale; 1 mark for 4 complete waves; 1 mark for correctly labelled axes.)
   b) i) Amplitude doubled and period halved; the coil is moving twice as fast and changing direction twice as often.
      ii) The sign of the output would be reversed; no change to amplitude or frequency.
      iii) Amplitude doubled because greater length of wire; no change to frequency.
2. a) The primary (coil) is the one receiving the input voltage; the secondary (coil) is the one delivering the output voltage.
   b) There are fewer turns on the secondary coil than on the primary coil.
   c) 143:1

Pages 96–97 Kinetic Theory

Multiple-choice questions
1. B 2. C 3. B 4. A 5. C

Short-answer questions
1. a) On the graph, the first slope should be labelled 'gas; the second slope should be labelled 'liquid'; and the third slope should be labelled 'solid'.
   b) i) 20°C
      ii) 120°C
   c) i) 293 K
      ii) 393 K
   d) 6.7 kJ

GCSE-style questions
1. a) Temperature
   b) Initial momentum = 0.25 kgm/s; change in momentum = 0.5 kgm/s
   c) i) A force on the ball is required to cause the change of direction; a reaction force is exerted on the wall.

ii) 5N
   d) The gas pressure is due to collision of molecules with the walls of the container.
   e) If the volume is reduced the molecules will collide with the wall more frequently; the pressure will be increased.
2. a) Evaporation is the gradual removal of water molecules that have sufficient energy to escape from the surface of the solution; boiling is a bulk change of state at a temperature where all molecules have sufficient energy.
   b) The energy required for evaporation comes from our bodies, thus causing a cooling effect.

Pages 98–99 The Gas Laws

Multiple-choice questions
1. B   2. C   3. A   4. A   5. A

Short-answer questions
1. a) The collisions of the gas molecules with the internal surface of the can and the surface of the liquid paint create the pressure.
   b) The pressure in the can drops; the paint comes out with less force.
2. a) More gas particles, means more collisions with the tyre wall.
   b) Temperature.

GCSE-style questions
1. a) i) A ii) B
   b) °C; if using K the graph should go through the origin.
2. a) Temperature decreases from 353 K to 293 K; therefore volume decreases in proportion to 415 ml.
   b) Temperature decreases in proportion $\frac{450}{500}$; to 318 K = 45°C
3. a) Pressure increases 4-fold; therefore temperature increases to 1172 K = 899°C.
   b) 293 k to 1473 k is a 5-fold increase; therefore pressure increases 5-fold to 500 kPa.

Pages 100–101 Medical Physics

Multiple-choice questions
1. B   2. C   3. D   4. C   5. A

Short-answer questions
1. tomography; tumours; glucose; radiopharmaceutical; annihilate; gamma; X-rays

GCSE-style questions
1. a) An anti-matter version of an electron; having the same mass but opposite charge.
   b) $^{18}_{9}F \longrightarrow {^{18}_{8}O} + {^{0}_{+1}\beta^+}$
   c) i) The isotope must be produced and prepared and injected into the patient very rapidly.
      ii) The isotope will not remain highly active for a long period meaning the patient will not be exposed to potentially damaging radiation for a prolonged period.
   d) i) The two particles destroy each other; their mass turns into energy.
      ii) To conserve momentum two gamma rays are emitted in opposite directions.
2. a) Total distance for sound to travel = 0.1 m; time = $\frac{\text{distance}}{\text{speed}}$ = 67 μs (6.7 × 10⁻⁵ s)
   b) Extra distance travelled = 0.04 m; time delay = 27 μs (2.7 × 10⁻⁵ s)

Answers

GCSE-style questions

Answer all parts of all questions. Continue on a separate sheet of paper if necessary.

1 A boy falls from a climbing frame to the ground in 0.7 s.

  a) Taking the value of acceleration due to gravity to be 10 m/s², calculate his velocity on impact. **(2 marks)**

  b) Calculate the momentum of the boy on impact if his mass is 40 kg. **(1 mark)**

  c) If the boy takes 0.2 s to come to rest, calculate the average force exerted on his body. **(2 marks)**

  d) Explain the advantage to the boy if the climbing frame had been installed over a safety surface that increases the time taken to break his fall to 0.5 s. **(2 marks)**

2 Wearing seatbelts in cars became compulsory for drivers and front seat passengers in 1992.

  a) A car of mass 1000 kg and driver of mass 70 kg travelling at 30 m/s collides with a stationary lorry and rapidly comes to rest. The car's crumple zone offers some protection by reducing the rate of deceleration. The vehicle takes 0.1 s to stop. Calculate the average force exerted on the car driver. **(3 marks)**

  b) State one other car safety design that reduces the rate of deceleration. **(1 mark)**

  c) Describe the evidence you would expect to support the safety benefit of either of the features mentioned that are designed to reduce the rate of change of momentum on the human body. Include in your answer one piece of data that would be usefully collected. **(2 marks)**

Score / 13

How well did you do?

0–6 Try again | 7–12 Getting there | 13–18 Good work | 19–24 Excellent!

For more information on this topic, see pages 62–63 of your Success Revision Guide.

Pairs of Forces: Action and Reaction

Multiple-choice questions

Choose just one answer: A, B, C or D.

① When two objects collide, the momentum afterwards: **(1 mark)**
- A is always more than before
- B is always the same as before
- C is always less than before
- D is always zero

② What is the 'reaction' to the 'action' of your weight? **(1 mark)**
- A a force pulling the Earth towards you
- B a force from the ground pushing back on you
- C the force you exert on the ground
- D the force of the Earth pulling you down

③ Two cars of mass 800 kg and 1000 kg have a head-on collision and come to a standstill. Calculate the speed of the second car, given that the first was moving at 10m/s **(1 mark)**

- A 10 m/s
- B 8 m/s
- C 5 m/s
- D 0 m/s

④ Which of the following is always conserved when objects collide or explode? **(1 mark)**
- A kinetic energy
- B velocity
- C momentum
- D acceleration

⑤ An orbiting spacecraft separates into two sections by means of booster rockets. If section A is half the mass of section B, which of the following is true? **(1 mark)**
- A the velocity of section B will double that of section A
- B the acceleration of each will be the same
- C the change in velocity of section A will be double that of section B
- D the acceleration of section B will be double that of section A

Score / 5

Short-answer questions

① Complete the following passage. **(10 marks)**

If two objects collide, they each have a change in their momentum because of their change in .... However, their combined momentum before and afterwards is ...., so we say that momentum is .... During the collision each object experiences a .... that causes an acceleration. The size of the force depends on how long it is acting: a large force acting for a .... time and a smaller force acting for a much .... time can cause the same change in momentum. The forces experienced by each object are the same .... because the rate of change of momentum is the same for each. However, they are in .... directions. The forces are called an .... and .... pair.

Score / 10

GCSE-style questions

Answer all parts of all questions. Continue on a separate sheet of paper if necessary.

1 The following diagram shows the Earth and the Moon.

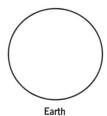

Earth   Moon

a) Using arrows, mark on the diagram the forces experienced by each body due to the gravitational attraction between them. **(3 marks)**

b) The forces are an action-reaction pair. Describe the relationship between these forces. **(2 marks)**

c) Below is a diagram showing a person standing on the surface of the Earth.

Indicate the action-reaction pair of forces due to the attraction between the Earth and the person. **(3 marks)**

2 A 1000 kg car and a 2500 kg truck are travelling at the same speed but in the opposite directions. The vehicles collide head-on.

a) Find an expression for the total momentum of the vehicles before they collide. **(3 marks)**

b) If the speed each vehicle was initially travelling at was 15 m/s, calculate the velocity of the wreckage immediately after the collision, assuming the two vehicles become entangled and move together as one object. **(4 marks)**

Score / 15

How well did you do?

| 0–7 Try again | 8–15 Getting there | 16–23 Good work | 24–30 Excellent! |

For more information on this topic, see pages 64–65 of your Success Revision Guide.

Work and Energy

Multiple-choice questions

Choose just one answer: A, B, C or D.

1 What work is done in lifting a textbook of mass 1.2 kg from the floor to a table at a height of 0.8 m? **(1 mark)**
- **A** 2 J
- **B** 1.5 J
- **C** 9.6 J
- **D** 15 J

2 Which of the following would not affect the work needed to be done against gravity to lift an object? **(1 mark)**
- **A** increasing the mass of the object
- **B** reducing the height difference
- **C** increasing the speed of lifting
- **D** lifting the object on the Moon instead of on the Earth

3 What is the kinetic energy of a 50 kg skateboarder moving at 20 m/s? **(1 mark)**
- **A** 10 kJ
- **B** 1 kJ
- **C** 25 kJ
- **D** 15 kJ

4 A ball of mass 100 g is thrown upwards with a velocity of 10 m/s. What is the maximum height it can reach? **(1 mark)**
- **A** 5 m
- **B** 10 m
- **C** 2.5 m
- **D** 0.5 m

5 An apple falls from a tree and hits the ground 2.5 m below with 3.5 J kinetic energy. What is the mass of the apple? **(1 mark)**
- **A** 8.8 kg
- **B** 140 g
- **C** 350 g
- **D** 1.4 kg

Score / 5

Short-answer questions

1 A crane raises a load of 500 kg from the ground to the top floor of a construction site at a height of 45 m.

a) Using the value of 10 N/kg for the gravitational field strength, calculate the change in gravitational potential energy (GPE) of the load. **(2 marks)**

b) On the grid alongside, draw a graph to show how the GPE of the load varies with its height during lifting to the top of the building. **(3 marks)**

c) Half-way up, the wire connected to the load snaps.

  i) Calculate the kinetic energy of the load when it hits the ground. **(2 marks)**

  ii) Calculate the velocity at which the load hits the ground. **(2 marks)**

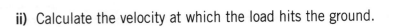

Score / 9

GCSE-style questions

Answer all parts of all questions. Continue on a separate sheet of paper if necessary.

1 The diagram shows a rollercoaster ride at a theme park. Initially, the roller coaster is winched mechanically to the starting point (1).

a) ✎ Explain the motion of the rollercoaster throughout the ride, referring appropriately to gravitational potential energy (GPE) and kinetic energy (KE) transfers at points 2–4. (Answer on a separate sheet) **(6 marks)**

b) Underline the correct answers to complete the following sentence. **(2 marks)**

The maximum height at any point of the ride is determined by **the speed at the lowest point / the starting height / the total mass** due to conservation of **momentum / mass / energy**.

c) The combined mass of the rollercoaster and its passengers is 2500 kg.

If the starting height is 40 m, calculate the KE of the roller coaster at point 3, when it is 5 m above the ground. Assume there are no energy losses due to friction. **(3 marks)**

d) Calculate how fast the passengers will be travelling when the rollercoaster passes point 3. **(2 marks)**

e) Without using any further calculations, describe how would your answer to **d)** would differ if an additional 10 passengers with a combined mass of 500 kg had boarded the rollercoaster at the start. **(3 marks)**

Score / 16

How well did you do?

| 0–8 Try again | 9–15 Getting there | 16–22 Good work | 23–30 Excellent! |

For more information on this topic, see pages 66–67 of your Success Revision Guide.

Energy and Power

Multiple-choice questions

Choose just one answer: A, B, C or D.

1. A man does 600 J of work in pushing a child in a toy truck for 15 m. What force does he exert? **(1 mark)**
   - A 400 N
   - B 40 N
   - C 4 N
   - D 9000 N

2. Select the correct definition of power. **(1 mark)**
   - A power is the amount or work done
   - B power is the rate of energy transfer
   - C power is the force exerted when doing work
   - D power is the maximum energy transfer

3. The Watt, the unit of power, is defined as a: **(1 mark)**
   - A J/s
   - B Nm
   - C kgm/s
   - D kgm/s^2

4. Which of the following does not affect thinking distance? **(1 mark)**
   - A speed
   - B tiredness of the driver
   - C condition of the brakes
   - D distraction of the driver

5. How much energy is transferred in 6 s at a power of 1.5 kW? **(1 mark)**
   - A 4 J
   - B 9000 J
   - C 250 J
   - D 4 kJ

Score / 5

Short-answer questions

1. A dog weighing 100 N runs up a hill that is 10 m high.

   a) Calculate the dog's energy transfer in working against gravity. **(1 mark)**

   b) If the dog reaches the top in 4 s, what is the power of the dog? **(1 mark)**

2. The following graph sketches how the speed of a skydiver changes with time from the moment he leaves the aircraft to just before opening the parachute.

   a) Continue the graph to show how the skydiver's speed changes for the remainder of the descent to Earth. **(3 marks)**

   b) Select the appropriate words to complete the following sentence. **(2 marks)**

   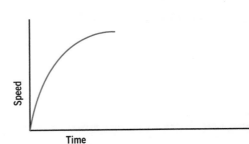

   At certain stages of the descent the forces acting on the skydiver are **unequal / balanced** so that the acceleration is zero. When this happens the skydiver is said to have reached **final / terminal** velocity.

Score / 7

GCSE-style questions

Answer all parts of all questions. Continue on a separate sheet of paper if necessary.

1 A 31 kg child climbs to the top of a 2.4 m slide and then slides down.

  a) Assuming the energy transfer from her gravitational potential energy is 100% efficient, calculate the average rate of kinetic energy transfer if the descent takes 1.2 s. **(3 marks)**

  b) If it took the child 6 s to climb up in the first place, what was the average power of her muscles working against gravity? **(2 marks)**

2 The driver of a 3 tonne lorry travelling at 80 km/h slams on the brakes when he sees a pedestrian step out into the road 100 m ahead.

  a) Calculate the kinetic energy of the lorry. **(2 marks)**

  b) Calculate the minimum braking force required if the lorry is to stop in time. Give your answer to 2 significant figures. **(2 marks)**

  c) In the event, the driver actually takes 0.9 s to react.

   i) Calculate the thinking distance. **(2 marks)**

   ii) Describe how this affects your answer to part **b)**. **(2 marks)**

  d) Select from the list provided one factor that affects each of the following: **(3 marks)**

   i) thinking distance only

   ii) braking distance only

   iii) both thinking and braking distance

   speed   driver's alertness   road conditions   brake or tyre condition   vehicle load

Score   / 16

How well did you do?

| 0–7 | Try again | 8–14 | Getting there | 15–21 | Good work | 22–28 | Excellent! |

For more information on this topic, see pages 68–69 of your Success Revision Guide.

Electrostatic Effects

Multiple-choice questions

Choose just one answer: A, B, C or D.

1 A charged plastic rod is suspended from a thread. A second charged rod, made from a different plastic, is brought near. The suspended rod is attracted to it. What can be concluded? **(1 mark)**
- **A** both rods are negatively charged
- **B** the suspended rod is negatively charged and the second rod is positive
- **C** the two rods have opposite charges
- **D** both rods are positively charged

2 What happens when a Perspex rod is positively charged by rubbing it on a cloth? **(1 mark)**
- **A** positive charges move from the cloth to the rod
- **B** electrons move from the cloth to the rod
- **C** positive and negative charges within the rod become separated at each end
- **D** electrons move from the rod to the cloth

3 A lightening conductor on a tall building works by: **(1 mark)**
- **A** preventing electrostatic charge build-up
- **B** providing a safe route for discharging
- **C** allowing the building to become charged
- **D** insulating the building from the effects of charge in the atmosphere

4 Static charge build-up poses a danger whilst refuelling aircraft. Why? **(1 mark)**
- **A** if the fuel becomes charged it will not burn efficiently in the aircraft engine
- **B** if a charge builds up sufficiently between the aircraft and the tanker there is a risk of discharge, causing an explosion
- **C** it will be more difficult to pump fuel if it has an opposite charge to the tanker
- **D** the charged fuel will charge the aircraft and affect the function of its instruments

Score / 4

Short-answer questions

1 When a balloon is rubbed on a jumper it becomes electrostatically charged. If it is then brought close enough it will be able to attract a small piece of paper, even though this is not charged.

Fill in the missing words to explain how the paper is attracted to the balloon. **(5 marks)**

.... transfer either from or to the ...., giving it a positive or negative charge. As the balloon is brought near the paper, the charges on the balloon induce a charge separation within the piece of paper. The charge on the part of the paper closest to the balloon is .... to the balloon's charge; these .... charges cause a force of ....

2 Conductors, such as a piece of metal, cannot be charged by rubbing in this way. What conditions are required to charge a conductor and why? **(2 marks)**

....

....

Score / 7

GCSE-style questions

Answer all parts of all questions. Continue on a separate sheet of paper if necessary.

1 a) An electroscope is a device that is used to show the presence of charge. If a charged object is placed on, or near to, the top plate then the gold leaf rises. This is because the stem of the electroscope and the gold leaf then become charged the same and the very thin gold leaf is repelled.

In the experiment shown, flour is being poured down a plastic chute into a metal can sitting on top of an electroscope. Explain how this models the potential build-up of electrostatic charge in a flour mill.

(6 marks)

b) A fine dispersion of flour particles in air is readily combustible. Describe how the presence of electrostatic charge in this environment poses a danger. **(2 marks)**

c) Explain what measure can be taken to avoid charge building up in the first place. **(2 marks)**

d) Identify one other situation in which electrostatic charge build-up poses a threat and describe what is done to minimise the danger in this situation. **(2 marks)**

Score / 12

How well did you do?

| 0–6 | Try again | 7–12 | Getting there | 13–18 | Good work | 19–23 | Excellent! |

For more information on this topic, see pages 72–73 of your Success Revision Guide.

Uses of Electrostatics

Multiple-choice questions

Choose just one answer: A, B, C or D.

1. Which of the following does NOT make use of electrostatic charge? **(1 mark)**
   - A removal of particulates from vehicle exhaust gases
   - B spray painting car bodies
   - C photocopiers
   - D heart defibrillation

2. Smoke particles can be removed from chimney exhausts by giving them a charge so that: **(1 mark)**
   - A the particles clump together
   - B the particles can be trapped on the grid
   - C the particles are repelled by the grid
   - D the particles can be attracted to the sides of the chimney

3. How do electrostatics aid crop spraying to control pests? **(1 mark)**
   - A by ensuring even coverage
   - B by activating the insecticide
   - C by attracting the insecticide to the pests
   - D by positively charging the plants that are sprayed

4. Positively charged paint droplets: **(1 mark)**
   - A have gained electrons
   - B have gained positive ions
   - C have lost electrons
   - D have lost positive ions

5. Which of the following is NOT an advantage of spray painting compared to other painting methods? **(1 mark)**
   - A paint is used more economically
   - B a greater variety of colour is available
   - C paint is applied evenly
   - D paint droplets repel each other

Score / 5

Short-answer questions

1. Complete the following passage. **(7 marks)**

   A bicycle frame is being prepared for spray painting. The frame is given a positive charge.

   The paint spray can is given a .... charge, so that the paint particles are

   also given this charge. Because the paint particles all have .... charge they

   .... each other to give a finely dispersed spray. The paint particles are

   .... by the bicycle frame because it has the .... charge.

   This reduces wastage of paint and also ensures that the coverage is ....,

   including on parts of the frame facing .... from the spray.

2. A heart defibrillator is charged by connecting it to a power supply. With reference to charge, describe how this device is used. **(3 marks)**
   (Answer on a separate sheet.)

Score / 10

GCSE-style questions

Answer all parts of all questions. Continue on a separate sheet of paper if necessary.

1 The following diagram shows how a laser printer works.

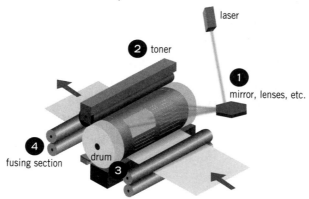

a) The following steps occur when a laser printer is used. Number each statement **1 to 6**, to show the correct sequence. **(5 marks)**

   A light selectively discharges some areas of the drum according to what is to be printed ☐

   B the rotating drum is charged up ☐

   C print toner is attracted to the charged areas only ☐

   D an image of what is to be printed is projected onto the rotating drum using laser light ☐

   E areas that are to be printed black have a different charge to those that will be white ☐

   F the print toner is attracted to the paper and fixed in place by heating ☐

b) What charge would be needed on the paper to help this process if the toner had a negative charge? Explain your answer. **(2 marks)**

2 Explain how static charge can be used to remove smoke particles from chimney exhausts. **(4 marks)**

Score / 11

How well did you do?

| 0–7 Try again | 8–14 Getting there | 15–20 Good work | 21–26 Excellent! |

For more information on this topic, see pages 74–75 of your Success Revision Guide.

Electric Circuits

Multiple-choice questions

Choose just one answer: A, B, C or D.

1 Which statement provides the best definition of electric current? **(1 mark)**
- A the rate of energy transfer
- B the rate of flow of charge
- C the flow of electrons
- D the amount of charge transfer

2 In a series circuit consisting of a battery and two bulbs, the current: **(1 mark)**
- A is highest for the bulb closest to the positive terminal of the battery
- B is shared equally between the two components, each having half the total current
- C is the same at all points of the circuit
- D is zero between the two bulbs

3 A circuit consists of a power pack and six parallel bulbs. What happens if one bulb fails? **(1 mark)**
- A the other bulbs all get a lot brighter
- B the other bulbs all go out
- C the other bulbs all get slightly dimmer
- D the other bulbs are unaffected

4 Which one of the following statements about alternating current in a lighting circuit is true? **(1 mark)**
- A the direction of charge flow regularly reverses
- B positive and negative charge carriers flow alternately
- C positive and negative charge carriers flow simultaneously
- D periods of charge flow alternate with periods without charge flow

Score / 4

Short-answer questions

1 Circle the correct options to complete the following explanation. **(8 marks)**

When a source of electrical energy, such as a battery, is connected into a circuit there will be a flow of **charge / gas / chemicals**, provided the circuit is **in series / continuous / parallel**. This flow is referred to as a **current / circuit / voltage**.

In a wire this involves the movement of **electrons / atoms / chemicals** but the flow is conventionally regarded as being from **positive / negative** to **negative / positive**. This flow is measured in units called **volts / amps / ohms**.

The flow at the positive terminus of the battery is **more than / the same as / less than** that at the negative terminus.

2 In a circuit, a 1.5 V battery provides a current of 750 mA to three identical bulbs connected in parallel.
Calculate the current for the following scenarios. **(6 marks)**

a) One bulb in the circuit fails _____

b) A 3 V battery is used instead _____

c) The bulbs are replaced with others rated as '1.5 V, 0.2 A' _____

Score / 14

GCSE-style questions

Answer all parts of all questions. Continue on a separate sheet of paper if necessary.

1 The diagram below shows three identical bulbs connected to a 3 V battery.

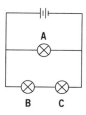

a) Add a switch (labelled X) to the above diagram that will control all three bulbs. **(1 mark)**

b) Add a second switch (Y) that will control Bulb A only. **(1 mark)**

c) Explain why it is not possible to use a switch to control Bulbs B and C independently. **(2 marks)**

d) An ammeter placed adjacent to Bulb A reads 0.2 A. Explain how you can use this information to deduce the reading on an ammeter placed adjacent to the battery (i.e. the total current supplied by the battery). Include the correct reading in your answer. **(4 marks)**

e) Draw another circuit, using the same components, in which all three bulbs have equivalent voltage and current readings to Bulb A. **(2 marks)**

f) Calculate the total current when a fourth equivalent bulb is added to your new circuit in parallel. **(2 marks)**

Score / 12

How well did you do?

| 0–8 | Try again | 9–14 | Getting there | 15–22 | Good work | 23–30 | Excellent! |

For more information on this topic, see pages 76–77 of your Success Revision Guide.

Voltage or Potential Difference

Multiple-choice questions

Choose just one answer: A, B, C or D.

① What does a voltmeter measure? **(1 mark)**
- A the flow of electricity
- B the electrical power
- C the electrical energy transfer in a component
- D the energy transfer per unit charge

② Which unit is used for charge? **(1 mark)**
- A amp
- B coulomb
- C joule
- D volt

③ How should an ammeter and a voltmeter be connected in a circuit? **(1 mark)**
- A both are connected in parallel
- B both are connected in series
- C the ammeter is connected in series; the voltmeter is connected in parallel
- D the voltmeter is connected in series; the ammeter is connected in parallel

④ What voltage is required to give 0.5 C of charge 4.5 J of energy? **(1 mark)**
- A 9 V
- B 5 V
- C 2.25 V
- D 0.11 V

⑤ Which statement is NOT true for a series circuit consisting of a 1.5 V cell, a switch and two identical bulbs? **(1 mark)**
- A with the switch closed, each bulb has a potential difference of 0.75 V
- B with the switch open, each bulb has a potential difference of 1.5 V
- C with the switch closed, the current is the same through each bulb
- D with the switch open, neither bulb is on because no current flows

Score / 5

Short-answer questions

① True or false? **True False (6 marks)**
- a) non-identical bulbs connected in series have the same current ☐ ☐
- b) identical bulbs connected in parallel have the same current ☐ ☐
- c) non-identical bulbs connected in series have the same potential difference ☐ ☐
- d) non-identical bulbs connected in parallel have the same potential difference ☐ ☐
- e) identical bulbs connected in series have the same potential difference ☐ ☐
- f) non-identical bulbs connected in parallel have the same current ☐ ☐

② For the circuit alongside, calculate the readings on the given ammeters and voltmeters. The bulbs are all identical. **(4 marks)**

- a) A_3 ....
- b) A_4 ....
- c) V_2 ....
- d) V_3 ....

Score / 10

GCSE-style questions

Answer all parts of all questions. Continue on a separate sheet of paper if necessary.

1 Twenty-five students were asked to build a circuit from the following components:
- two identical bulbs
- one 3 V battery
- voltmeters to measure the potential difference across each bulb

a) On a separate piece of paper, draw each of the different types of circuit that the students could have built. Take care to use the correct symbols. (4 marks)

b) How does the brightness of the bulbs in each circuit compare? Explain your answer. (3 marks)

c) The results obtained by the students are grouped in the table below:

Group	Number of students	Bulb X	Bulb Y
A	14	1.5 V	1.5 V
B	10	3.0 V	3.0 V
C	1	1.5 V	3.0 V

   i) What might explain the results obtained by the student in Group C? (2 marks)

   ii) Suggest how you could test your answer to part **i)** (1 mark)

d) State the type of circuit built by each of the following groups: (2 marks)

   i) Group A: _____

   ii) Group B: _____

e) Calculate the current in the circuit built by Group A, if the energy transfer in each bulb is 0.3 J every second. (3 marks)

Score / 15

How well did you do?

| 0–8 | Try again | 9–14 | Getting there | 15–22 | Good work | 23–30 | Excellent! |

For more information on this topic, see pages 78–79 of your Success Revision Guide.

Resistance and Resistors

Multiple-choice questions

Choose just one answer: A, B, C or D.

1 Resistance in a metal is due to: (1 mark)
   A collisions of electrons with the lattice ions
   B restricted movement of the lattice ions
   C the voltage applied
   D freely moving electrons

2 A component whose resistance has a fixed value for a range of potential differences is said to obey which law? (1 mark)
   A Coulomb's Law
   B Newton's Law
   C Ohm's Law
   D the law of conservation of energy

3 Which of the following is NOT true for a fixed resistor? (1 mark)
   A doubling the voltage doubles the current
   B the combined resistance of two parallel resistors is half that of one alone
   C putting two such resistors in series has no effect on the current for a given voltage
   D halving the current halves the voltage

4 What is the combined resistance of three 10 Ω resistors in series? (1 mark)
   A 1000 Ω   B 30 Ω
   C 3.3 Ω   D 10 Ω

5 Resistors X and Y are connected in series to a 3 V battery. A current of 0.2 A flows through resistor X. A potential difference of 2 V is measured for resistor Y. What are the values of X and Y? (1 mark)
   A X=10 Ω; Y=5 Ω   B X=15 Ω; Y=5 Ω
   C X=6 Ω; Y=12 Ω   D X=5 Ω; Y=10 Ω

Score / 5

Short-answer questions

1 Metals are thought of as a lattice of stationary positive ions surrounded by free electrons.

   a) State which feature of this structure explains the electrical conductivity of metals. (1 mark)

   b) Describe what happens if a potential difference is applied across the piece of metal. (1 mark)

   c) What is different about the structure and behaviour of an insulator compared to metals? (1 mark)

2 A fixed resistance of 10 Ω is connected to a 3 V battery.

   a) Calculate the current that flows. (1 mark)

   b) What would be the effect of adding a second, identical resistor in series? (1 mark)

   c) What would be the effect of adding a second, identical resistor in parallel? (1 mark)

Score / 6

GCSE-style questions

Answer all parts of all questions. Continue on a separate sheet of paper if necessary.

1 Resistance can be thought of as a 'force' opposing a current.

  a) State which feature of the structure of metals causes resistance? **(1 mark)**

  b) The potential difference across a resistor indicates the difference in energy per unit charge flowing into and out of the resistor. Explain what accounts for this energy difference. **(2 marks)**

  c) Describe how resistance is defined. Remember to state the unit used. **(2 marks)**

2 A bulb with resistance 8 Ω is connected in series with a second bulb of 16 Ω. Both bulbs are then connected to a 3 V battery.

  a) Calculate the reading on an ammeter placed between the bulbs. **(1 mark)**

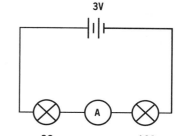

  b) Calculate the potential difference across each of the bulbs. **(2 marks)**

  c) i) What would be the current through each bulb if they were connected in parallel to the battery? **(2 marks)**

  ii) Use your answer to part **i)** to calculate the total current from the battery and, therefore, the effective resistance of the two bulbs together. **(2 marks)**

Score / 12

How well did you do?

0–6 Try again | 7–12 Getting there | 13–18 Good work | 19–23 Excellent!

Special Resistors

Multiple-choice questions

Choose just one answer: A, B, C or D.

1. Which of the following resistors could be used to control a thermostat? **(1 mark)**
   - A light-dependent resistor
   - B diode
   - C thermistor
   - D light-emitting diode

2. Which of the following does not make use of a variable resistor? **(1 mark)**
   - A the volume control on a radio
   - B a dimmer switch for a light
   - C the 'charging' light on a rechargeable device
   - D the shutter control on a camera

3. Which of the following components shows a strong dependence on polarity (i.e. its orientation in the circuit)? **(1 mark)**
   - A diode
   - B thermistor
   - C LDR
   - D variable resistor

4. The initials NTC are applied to a component called a thermistor. What do the initials stand for? **(1 mark)**
   - A non-thermal conductor
   - B negative terminal charge
   - C negative temperature coefficient
   - D new thermal component

5. A variable resistor was used to set the current in a circuit to 0.3 A. What resistance did it have if its p.d. was 1.5 V? **(1 mark)**
   - A 0.45 Ω
   - B 0.2 Ω
   - C 5 Ω
   - D 1.5 Ω

Score / 5

Short-answer questions

1. During its operation a filament bulb gets hot.

   a) Explain how temperature affects resistance in this case. **(2 marks)**

   b) Describe how your answer to part **a)** differs from the behaviour of a thermistor. **(1 mark)**

   c) The grid below displays a graph of the dependence of current on voltage for a thermistor. Add a second graph showing what you would expect if you were to do the same experiment at a higher temperature. **(1 mark)**

   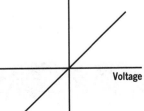

Score / 4

GCSE-style questions

Answer all parts of all questions. Continue on a separate sheet of paper if necessary.

1 The following graph shows how the current through a filament bulb varies with applied voltage.

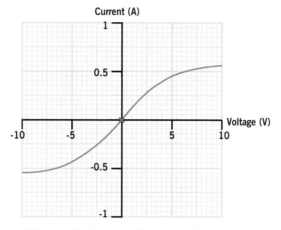

a) Calculate the resistance of the bulb for the given applied voltages: **(2 marks)**

   i) 2 V ....

   ii) 6 V ....

b) Would you describe the bulb as an Ohmic or non-Ohmic conductor? Explain your answer. **(2 marks)**

c) On the axes below, sketch the graph you might expect if you were to carry out the same experiment using a diode instead of the bulb. **(3 marks)**

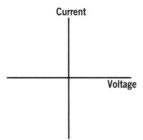

d) Describe the role of a diode in a circuit. **(1 mark)**

2 Describe the resistance behaviour shown by a component known as a 'LDR'. **(2 marks)**

Score / 10

How well did you do?

0–5 Try again | 6–10 Getting there | 11–15 Good work | 16–19 Excellent!

For more information on this topic, see pages 82–83 of your Success Revision Guide.

The Mains Supply

Multiple-choice questions

Choose just one answer: A, B, C or D.

1 What colour is the live wire in the mains electricity supply? **(1 mark)**
- A red
- B blue
- C brown
- D green and yellow

2 Which of the following best describes mains electricity? **(1 mark)**
- A 230 Hz a.c., 50 V
- B 50 Hz a.c., 230 V
- C 100 Hz a.c., 230 V
- D 60 Hz a.c., 115 V

3 Which of the following is NOT a safety feature for use with mains electricity? **(1 mark)**
- A RCCB
- B use of alternating current
- C double insulation
- D earthing

4 A mains-operated television requires 0.25 A. What is its approximate power rating? **(1 mark)**
- A 60 W
- B 1k W
- C 1m W
- D 15 W

5 A device is rated at using 4.2 A. What fuse rating is most appropriate? **(1 mark)**
- A 5 A
- B 3 A
- C 13 A
- D 1 A

Score / 5

Short-answer questions

1 The diagram shows the inside of a mains plug.

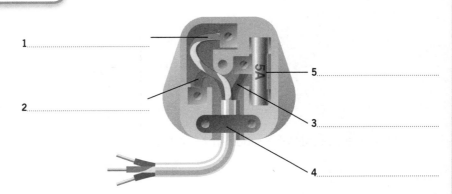

  a) Add labels for the indicated parts of a mains plug. **(5 marks)**

  b) Explain the function of component 5 and how it works. **(2 marks)**

  c) The maximum current rating of a mains-operated device is 13 A. Calculate the maximum power of such a device, giving the answer to 2 significant figures in kW. **(2 marks)**

Score / 9

GCSE-style questions

Answer all parts of all questions. Continue on a separate sheet of paper if necessary.

1 A key safety feature of the mains electricity supply is the inclusion of an earth wire. A device that has an external case made entirely of plastic, such as a hair-drier, does not need an earth connection.

a) Explain what it means to 'earth' something, indicating how this protects the user of the equipment in the event of an electrical fault. **(3 marks)**

b) Describe why a device with an external case made entirely of plastic, does not need an earth connection. You should include the correct term for this type of protection in your answer. **(2 marks)**

2 The picture opposite shows a residual current circuit breaker (RCCB) that is recommended for use with power tools. The RCCB plugs into the socket before the power tool is plugged into the RCCB.

a) Describe what is meant by 'residual current'? **(1 mark)**

b) Describe how the RCCB offers protection to the person using the power tool. **(2 marks)**

c) Explain why using a circuit breaker is better than relying on just the fuse. **(2 marks)**

Score / 10

How well did you do?

| 0–6 | Try again | 7–12 | Getting there | 13–18 | Good work | 19–24 | Excellent! |

For more information on this topic, see pages 84–85 of your Success Revision Guide.

Atomic Structure

Multiple-choice questions

Choose just one answer: A, B, C or D.

1 Which of the following most accurately describes atomic structure? **(1 mark)**
   A the atom is made of positively charged material with embedded electrons
   B neutrons orbit a nucleus composed of protons and electrons
   C electrons orbit a nucleus composed of protons and neutrons
   D proton-electron pairs orbit a nucleus composed of neutrons

2 In the symbols used to represent an atom, Z represents the number of: **(1 mark)**
   A protons
   B neutrons
   C charges
   D electrons

3 An isotope of carbon has 6 protons and 6 neutrons. A second isotope has 8 neutrons; what is its mass number? **(1 mark)**
   A 6   B 8
   C 12   D 14

4 What are protons and neutrons collectively known as? **(1 mark)**
   A isotopes
   B ions
   C atoms
   D nucleons

5 During nuclear decay, which of the following may NOT be conserved? **(1 mark)**
   A charge
   B number of protons
   C atomic mass
   D number of nucleons

Score / 5

Short-answer questions

1 Our modern view of the atom involves particles known as electrons, protons and neutrons.
   a) Complete the following table of properties of these particles. **(4 marks)**

Particle	Relative mass	Charge
Electron	$\frac{1}{1840}$	
Proton		+1
Neutron		

   b) State the relationship between the number of electrons and the number of protons for any given neutral atom. **(1 mark)**

   ....

2 How many protons, neutrons and electrons do the following have? **(3 marks)**
   a) neutral atom of $^{56}_{26}Fe$ ....
   b) a neutral atom of $^{32}_{15}P$ ....
   c) a K+ ion of $^{39}_{19}K$ ....

Score / 8

GCSE-style questions

Answer all parts of all questions. Continue on a separate sheet of paper if necessary.

1 As knowledge of atomic structure has developed, scientists have revised the working model of the atom.

   a) With the aid of a diagram, describe the model devised by J.J. Thompson, usually referred to as the 'plum pudding' model. **(3 marks)**

   b) A classic experiment by Geiger and Marsden led Ernest Rutherford to propose an entirely different model of the atom. The surprising finding was that alpha particles were occasionally deflected or bounced back when fired at a very thin piece of gold foil, though most went straight through. From the following list select two conclusions that Rutherford drew from these observations. Tick the correct statements. **(2 marks)**

   A Atomic nuclei contain neutrons. ☐

   B Alpha particles are positively charged. ☐

   C The mass of the atom is concentrated in a very small region. ☐

   D Gold atoms are positively charged. ☐

   E Most of the atom is empty space. ☐

2 A text book refers to the following isotopes of oxygen:

$$^{16}_{8}O \quad ^{17}_{7}O \quad ^{17}_{8}O \quad ^{18}_{8}O$$

   a) Explain why at least one of these must be incorrect. **(2 marks)**

   b) Describe what is meant by the term 'isotope'. **(2 marks)**

3 Some atoms are unstable and undergo radioactive decay. Complete the following alpha decay equation for radon: **(2 marks)**

$$^{222}_{86}Rn \rightarrow ^{\Box}_{\Box}Po + ^{4}_{2}He$$

Score / 11

How well did you do?

| 0–6 Try again | 7–12 Getting there | 13–18 Good work | 19–24 Excellent! |

For more information on this topic, see pages 88–89 of your Success Revision Guide.

Radioactive Decay

Multiple-choice questions

Choose just one answer: A, B, C or D.

1 What is a beta particle? **(1 mark)**
  A an electron from the atom
  B a helium nucleus
  C a fast moving electron emitted by the nucleus
  D high frequency electromagnetic radiation

2 What is gamma radiation effectively stopped by? **(1 mark)**
  A the outer layer of skin
  B 5 mm thickness of lead
  C several metres of concrete
  D 5 mm thickness of aluminium

3 A radioisotope sample has 8×10^9 undecayed nuclei. How many will remain 1 hour later if the half-life is 15 min? **(1 mark)**
  A 2×10^9
  B 5×10^8
  C 2.5×10^8
  D 1×10^9

4 Carbon-14 is a radioactive isotope, $^{14}_{6}C$. Which of the following represents a Carbon-14 atom following beta decay? **(1 mark)**
  A $^{15}_{6}N$
  B $^{15}_{7}N$
  C $^{13}_{5}N$
  D $^{14}_{7}N$

5 During gamma decay: **(1 mark)**
  A there is no change to the nucleus
  B the nucleus relaxes to a lower energy state
  C nuclear particles disintegrate
  D radio waves are emitted

Score / 5

Short-answer questions

1 True or false? True False **(6 marks)**
  a) All atoms of all elements undergo radioactive decay.
  b) Alpha particles have the least ability to penetrate matter.
  c) When atoms undergo gamma decay the element changes.
  d) Particles released by radioactive decay can cause ionisation of matter.
  e) A sheet of paper is sufficient to stop alpha particles.
  f) When atoms decay there is always a reduction in the atomic number.

2 A sample of a radioisotope is monitored and it is found that there are 4.8×10^5 disintegrations in 2 minutes.
  a) Ring the option below that shows the activity of this sample. **(1 mark)**

  4000 Bq   400 Bq   480 000 Bq   240 000 Bq   8000 Bq

  b) If the half-life of the radioisotope is 2.5 h, calculate the activity expected after 7.5 h. **(2 marks)**

Score / 9

GCSE-style questions

Answer all parts of all questions. Continue on a separate sheet of paper if necessary.

1 Sam is a research scientist. He is asked to investigate an isotope of bromine that is found to be radioactive. Sam carries out an experiment where he places different materials between a sample of the bromine isotope and a detector.

a) Describe the results that you would expect Sam to get if the material was emitting both beta and gamma radiation. **(3 marks)**

b) Describe what safety precautions Sam should take when carrying out this experiment. **(2 marks)**

c) Sam reads that the radioactive decay of the bromine isotope has a 'half-life' of 2.4 hours. State what is meant by the term 'half-life'. **(1 mark)**

d) Sam is confused to find another book that says 'it is impossible to say how long it will be before any individual atom decays.' Explain how both the comments read by Sam can be true. **(6 marks)**

e) A colleague tells Sam that heating the bromine isotope will increase the rate of decay. Is this true or false? **(1 mark)**

Score / 13

How well did you do?

| 0–8 | Try again | 9–14 | Getting there | 15–19 | Good work | 20–27 | Excellent! |

For more information on this topic, see pages 90–91 of your Success Revision Guide.

Living with Radioactivity

Multiple-choice questions

Choose just one answer: A, B, C or D.

1 Identify the correct statement. **(1 mark)**
- A all types of radiation are equally dangerous
- B the majority of background exposure to radiation is natural
- C gamma rays are the most energetic and therefore most ionising
- D alpha particles are the most massive and therefore most difficult to stop

2 A radioactive leak consists of a mixture of substances with different properties. Which of the following presents the greatest long-term problem? **(1 mark)**
- A high activity, short half-life, chemically inert
- B low activity, long half-life, chemically inert
- C low activity, long half-life, readily incorporated into biomolecules
- D high activity, short half-life, readily incorporated into biomolecules

3 Which of the following is the greatest source of background radiation? **(1 mark)**
- A rocks and buildings
- B nuclear power
- C food and drink
- D radon gas

4 Which of the following would block beta but not gamma radiation? **(1 mark)**
- A paper
- B 50 mm lead
- C clothing
- D 5 mm aluminium

5 People who fly frequently are at risk from increased exposure to what type of background radiation? **(1 mark)**
- A radon gas
- B cosmic rays
- C consumption of irradiated food
- D radioactive waste

Score / 5

Short-answer questions

1 a) The following statements are true:
- One source of the background radiation exposure we receive is from our food.
- Modern food industry practice includes irradiation of some foods in order to increase their shelf-life. The conclusion that food irradiation contributes to our background exposure is, however, a misconception. Explain why. **(2 marks)**

b) Following the Chernobyl incident of 1986, grass used for grazing sheep in Britain was contaminated with radioactive material. To protect the public, these sheep were not used for food. Why are the consequences of contamination of the food chain of serious concern in terms of radiation exposure to our bodies? **(2 marks)**

Score / 4

GCSE-style questions

Answer all parts of all questions. Continue on a separate sheet of paper if necessary.

1 The following chart indicates the different sources of background radiation.

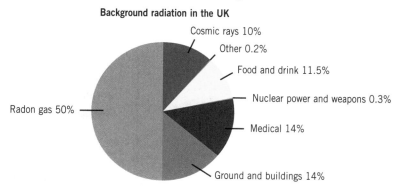

a) Complete the sentence: 'Background radiation refers to...' **(1 mark)**

b) Given that radon is a gas that is itself radioactive, explain why special precautions are needed for buildings built upon granite. **(2 marks)**

c) ✏ The decay products of radon are not gases, but solids. Although external exposure to alpha radiation is not generally hazardous, what are the possible consequences for health of inhalation of radon? **(6 marks)**
(Answer on a separate sheet)

2 A scientist has a small spillage of a radioactive material in the laboratory. He uses a Geiger counter to monitor the area, obtaining a reading of 212 Bq. With a half-life of 110 min, she calculates that the activity will have decayed effectively to zero by the following day and so is surprised to find an activity of approximately 2 Bq when she checks the area.

a) Describe what calculation the scientist used to justify thinking that the material would have decayed within 24 hours. **(2 marks)**

b) Describe the basic experimental error that she made that accounts for the remaining count of 2 Bq. **(2 marks)**

Score / 13

How well did you do?

0–6 Try again 7–11 Getting there 12–16 Good work 17–22 Excellent!

For more information on this topic, see pages 92–93 of your Success Revision Guide.

Uses of Radioactive Materials

Multiple-choice questions

Choose just one answer: A, B, C or D.

1. What kind of radiation is used for sterilising medical equipment or increasing the shelf-life of food? **(1 mark)**
   - A alpha
   - B beta
   - C gamma
   - D X-ray

2. Radiation workers must avoid occupational exposure as much as possible. Which of the following is NOT a means of reducing exposure? **(1 mark)**
   - A monitoring
   - B reducing contact time
   - C using shielding
   - D maximising their distance

3. In which of the following processes is radioactivity NOT involved? **(1 mark)**
   - A locating blockages in underground pipes
   - B paper milling
   - C cancer diagnosis
   - D laser printing

4. In igneous rocks, lead is a stable product formed by the decay of uranium radioisotopes. The ratio of lead to uranium can be used for dating. If the ratio is 3:1, how many half-lives have elapsed since the rock solidified? **(1 mark)**
   - A 1
   - B 2
   - C 3
   - D 4

5. Material that once was living can be dated using a radioisotope of which element? **(1 mark)**
   - A oxygen
   - B potassium
   - C carbon
   - D nitrogen

Score / 5

Short-answer questions

1. The radioisotope americium-241 is used in smoke detectors.

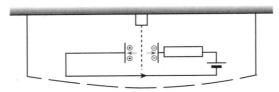

   With reference to how the smoke alarm works, explain why beta and gamma radiation could not be used in the same way. **(2 marks)**

2. An underground pipe is suspected to have a leak somewhere in a 100 m stretch. A radioisotope tracer could be used to locate and patch the leak. State what type of emission is needed for this technique and explain your choice. **(2 marks)**

Score / 4

GCSE-style questions

Answer all parts of all questions. Continue on a separate sheet of paper if necessary.

1 Aluminium foil production can be monitored using a radioactive source and detector in order to standardise thickness, as shown in the diagram.

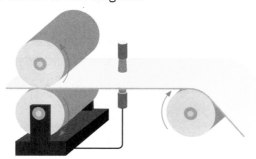

a) Describe how measuring the radioactivity passing through the foil be used to judge thickness. **(1 mark)**

b) State the type of emission that would be most appropriate. **(1 mark)**

c) State if your answer to **b)** would differ if the material being produced was sheet steel and explain your answer. **(2 marks)**

d) Would an isotope of long or short half-life be most advantageous for this type of industrial process? Explain your answer. **(2 marks)**

2 The radioisotope technetium-99 m, which is used for bone scans, decays by gamma emission with a half-life of 6 h. It is linked to a phosphate derivative and injected into the patient, where it gets incorporated at sites of bone growth. A camera can then be used to scan the patient for sites of accumulation – for example, to aid diagnosis of bone cancers. Explain, by referring to patient risk and patient benefit, why this technique would be justified to aid management of a cancer patient. **(2 marks)**

Score / 8

How well did you do?

| 0–4 Try again | 5–9 Getting there | 10–13 Good work | 14–17 Excellent! |

For more information on this topic, see pages 94–95 of your Success Revision Guide.

Nuclear Fission and Fusion

Multiple-choice questions

Choose just one answer: A, B, C or D.

① The energy-releasing process that occurs in stars is: **(1 mark)**
- A fission
- B fissile
- C fusion
- D futile

② Which of the following statements is true regarding the renewability of nuclear power? **(1 mark)**
- A fusion is renewable but fission is not
- B fission is renewable but fusion is not
- C both fission and fusion are renewable
- D neither fission nor fusion are renewable

③ A self-sustaining series of fission events is called a: **(1 mark)**
- A domino effect
- B neutron shower
- C chain reaction
- D landslide reaction

④ In a fission reactor the purpose of the moderator is: **(1 mark)**
- A to prevent a chain reaction
- B to slow down the neutrons to encourage a chain reaction
- C to increase the energy output
- D to reduce radioactive waste products

⑤ A major environmental advantage of nuclear power is: **(1 mark)**
- A many jobs are created
- B no pollutant gases are released
- C the energy released per kg of fuel is very high
- D power stations require a plentiful water supply

Score / 5

Short-answer questions

① a) Fill in the missing words to complete these statements about nuclear reactors. **(6 marks)**

i) The .... are made of uranium-235 or plutonium-239.

ii) A .... is used to slow down the neutrons so that they can be absorbed.

iii) The energy heats up the ....

iv) A .... is circulated to remove the heat.

v) The hot coolant is used to produce steam, which drives the power station's ....

vi) Control rods can be moved into the reactor to absorb .... and slow or stop the reaction.

b) Draw lines to match the different levels of radioactive waste to the disposal method used. **(2 marks)**

Low level waste	Kept under water in cooling tanks
Intermediate level waste	Mixed with concrete and stored in stainless steel containers.
High level waste	Sealed into containers and put in landfill sites.

Score / 8

GCSE-style questions

Answer all parts of all questions. Continue on a separate sheet of paper if necessary.

1 The following nuclear equation describes the process of nuclear fusion, which occurs in stars. It is responsible for releasing large amounts of electromagnetic radiation.

$$^{2}_{1}H + ^{1}_{1}H \rightarrow ^{3}_{2}He$$

a) State what type of force must be overcome in order to allow the nuclear strong force to take over and join the nuclei together. **(1 mark)**

b) Scientists are working to try and create conditions that would allow fusion to occur as a means of generating electricity.

   i) State the two extreme conditions that are needed, which scientists are struggling to achieve. **(2 marks)**

   ii) State TWO advantages that this process has over the existing process of nuclear power generation. **(2 marks)**

2 When U-235 undergoes fission two new nuclei are formed, such as Barium-141 and Krypton-92.

a) Complete the following decay equation: **(2 marks)**

$$^{235}_{92}U + ^{1}_{0}n \rightarrow ^{\square}_{56}Ba + ^{92}_{\square}Kr + 3^{1}_{0}n$$

b) An alternative pair of daughter nuclei is Xenon-140 (Z = 54) and Strontium-94 (Z = 38). Calculate how many neutrons are released by this fission event. **(1 mark)**

c) Write a decay equation for the fission event in part **b)**. **(3 marks)**

Score / 11

How well did you do?

| 0–6 Try again | 7–12 Getting there | 13–18 Good work | 19–24 Excellent! |

For more information on this topic, see pages 96–97 of your Success Revision Guide.

Refraction, Dispersion and TIR

Multiple-choice questions

Choose just one answer: A, B, C or D.

1. Why does light change direction when leaving a glass block? **(1 mark)**
   - A because of a change of frequency
   - B because it travels more slowly in air
   - C because it travels faster in air
   - D because it gets diffracted

2. Which of the following is not an application of total internal reflection in optical fibres? **(1 mark)**
   - A transatlantic communication
   - B keyhole surgery in medicine
   - C radio communication
   - D computer networking

3. The colours of the rainbow are dispersed when light enters and leaves droplets of rainwater. What is this process called? **(1 mark)**
   - A refraction
   - B diffraction
   - C reflection
   - D polarisation

4. When light enters water from air at an angle, which direction is it refracted in? **(1 mark)**
   - A away from the normal
   - B towards the surface
   - C towards the normal
   - D away from the surface

5. What is an endoscope used for? **(1 mark)**
   - A to see inside confined spaces
   - B to see around obstacles
   - C to measure the frequency of alternating voltage
   - D to measure the angle of refraction of light

Score / 5

Short-answer questions

1. Refraction of light refers to a change in its direction when leaving one medium and entering another.

   a) What does the term 'medium' refer to in this context? **(1 mark)**

   b) What two conditions are necessary for a change of direction to occur? **(2 marks)**

2. The diagram below shows a glass prism and a ray of visible light.

   a) Continue the ray diagram to show the path taken by (i) red light and (ii) blue light. **(4 marks)**

   b) State what you would see if the incident ray was white light. **(2 marks)**

Score / 9

GCSE-style questions

Answer all parts of all questions. Continue on a separate sheet of paper if necessary.

1 The diagram shows a light ray leaving a Perspex block and entering air.

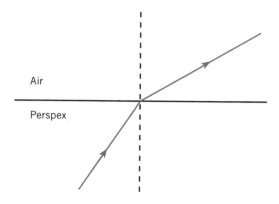

Snell's Law gives the relationship between the angles of incidence and refraction depending on the refractive index of the medium compared to a vacuum. In this case $n_{Perspex} = \frac{\sin(r)}{\sin(i)}$.

a) On the diagram mark the angles of incidence (*i*) and refraction (*r*). (2 marks)

b) Measure the following angles: (2 marks)

  i) value of *i* ....

  ii) value of *r* ....

  iii) Use your answers to parts **i)** and **ii)** to show that the refractive index ($n_{Perspex}$) is approximately 1.5 (1 mark)

c) i) What does this value for the refractive index indicate about the speed of light in Perspex? Tick the correct option. (1 mark)

  A It is lower than the speed in air. ☐

  B It is the same as the speed in air. ☐

  C It is greater than the speed in air. ☐

  ii) Using a value for $n = 1.5$ and given that the speed of light in air is 3×10^8 m/s, calculate the speed of light in Perspex. (1 mark)

d) If $n = 1.5$, calculate the angle of refraction for an incident angle of 40°. (2 marks)

e) Calculate the critical angle for Perspex (the angle for which light refracts at 90°). (2 marks)

f) What phenomenon occurs for incident angles greater than the critical angle? (1 mark)

Score / 12

How well did you do?

0–6 Try again 7–13 Getting there 14–20 Good work 21–26 Excellent!

For more information on this topic, see pages 100–101 of your Success Revision Guide.

Lenses

Multiple-choice questions

Choose just one answer: A, B, C or D.

1 Light emitted towards a lens from a point at the focal distance will: (1 mark)
- A emerge as a parallel beam
- B converge towards the focal point on the other side of the lens
- C diverge more strongly than without the lens
- D be internally reflected within the lens

2 An object viewed at a distance less than the focal distance of a convex lens will be: (1 mark)
- A inverted
- B real
- C distorted
- D magnified

3 A lens has a power of −2D. What does the minus sign indicate? (1 mark)
- A the image is inverted
- B the image is reduced
- C the lens is very weak
- D it is a diverging lens

4 A lens focuses parallel light at a distance of 50 cm. What is the power of this lens? (1 mark)
- A 0.02 D
- B 0.5 D
- C 2 D
- D 50 D

5 A camera lens produces an image of a 1.5 m boy that is 22.5 mm tall. What is the magnification? (1 mark)
- A 0.015
- B 67
- C 0.067
- D 15

Score / 5

Short-answer questions

1 The ray diagram below shows a lens and a tree. F is the position of the focal point on either side of the lens.

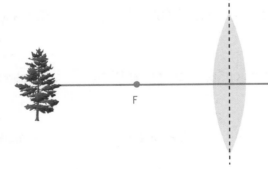

a) Add lines to complete the ray diagram, to indicate the position at which a screen would need to be placed to form a focussed image of the tree. (3 marks)

b) Circle the correct option to complete the following statements and describe the image produced: (2 marks)

   i) The image is **real** / **virtual**.

   ii) The image is **upright** / **inverted**.

Score / 5

GCSE-style questions

Answer all parts of all questions. Continue on a separate sheet of paper if necessary.

1 A student arranges a number of convex lenses in order of increasing thickness, expecting that the power of the lens will increase in the same order.

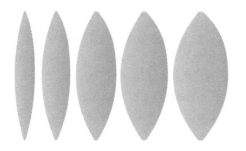

a) State what is meant by the 'power' of a lens. **(1 mark)**

b) Describe how the power of a lens can be determined experimentally. **(3 marks)**

c) The student is surprised to find that the power of the lenses does not increase in the same order as thickness of the lens. Of the situations described below, which two could explain why the results are not as the student expected? Tick the correct options. **(2 marks)**

  A The refractive index of the lens material is different for all or some of the lenses.

  B The refractive index of the lens material is the same for all the lenses.

  C The angle of the curvature of the lens does not increase in direct proportion to the thickness of the lens.

  D The angle of curvature is the same for all the lenses.

2 When white light is refracted by glass the component colours are separated.

a) Complete the following sentence. This phenomenon is known as ....; it occurs because the angle of refraction of glass is dependent on .... **(2 marks)**

b) Light from a distant source is brought to a focus. Explain which end of the visible spectrum (red/blue) will be brought to a focus closest to the lens. **(2 marks)**

Score / 10

How well did you do?

| 0–5 Try again | 6–10 Getting there | 11–15 Good work | 16–20 Excellent! |

For more information on this topic, see pages 102–103 of your Success Revision Guide.

Seeing Images

Multiple-choice questions

Choose just one answer: A, B, C or D.

1 A distance of 25 cm is taken to be the closest point at which the human eye can focus; infinity is taken to be the furthest. What are these positions known as? **(1 mark)**
- A the closest point/furthest point
- B the near point/far point
- C the short point/long point
- D the earliest point/latest point

2 What kind of lens would be used to correct the eyesight of someone with mild long-sightedness as they age? **(1 mark)**
- A strongly converging
- B strongly diverging
- C weakly converging
- D weakly diverging

3 What type of image is formed by a concave lens? **(1 mark)**
- A real, upright and magnified
- B virtual, inverted and reduced
- C virtual, upright and reduced
- D virtual, upright and magnified

4 What measurement do you obtain by dividing image height by object height? **(1 mark)**
- A magnification
- B power of the lens
- C focal distance
- D aberration

5 Which of the following is a possible cause of short-sightedness? **(1 mark)**
- A the cornea is too flat
- B the lens is too weak
- C the eyeball is too long
- D the eyeball is too short

Score / 5

Short-answer questions

1 The human eye is capable of focussing light coming from objects at a vast range of distances, from approximately 25 cm (e.g. when reading) to infinite distances (e.g. when looking at the stars).

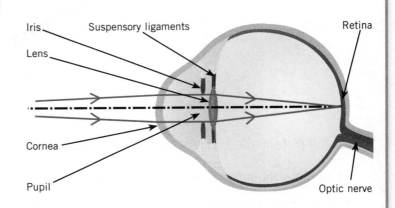

a) State which two features enable the eye to focus light. **(2 marks)**

b) The ability to switch focus when viewing objects at different distances is called accommodation. In which of the following situations does the eye have the greatest power? Tick the correct option. **(1 mark)**

- A The eye focused on a near object, e.g. a book or magazine. ☐
- B The eye focused on a distant object, e.g. a tree on the horizon. ☐

Score / 3

GCSE-style questions

Answer all parts of all questions. Continue on a separate sheet of paper if necessary.

1 In the diagram below the arrow represents a page of small print. The diagram shows how a convex lens can be used as a magnifying glass to enlarge the print. F is the focal point on either side of the lens.

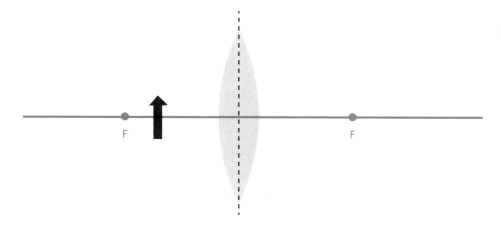

a) On the diagram, add lines to construct a ray diagram and show how the lens magnifies the print. Identify the positions of both the top and bottom of the image. **(5 marks)**

b) State the type of image produced and its orientation. **(2 marks)**

c) From the diagram, estimate the factor by which the image is magnified. **(1 mark)**

2 Similar to the human eye, a camera can be adjusted to focus near and far. Unlike the eye, however, the focussing power is not changed. Describe what method is used by the camera to adjust the position at which the image is in focus. **(2 marks)**

3 ✏ Someone who is short-sighted is unable to focus light from a distance on the retina. Explain short-sightedness in terms of power and describe how a lens could be used to correct this person's sight. **(6 marks)**

Score / 16

How well did you do?

| 0–6 | Try again | 7–12 | Getting there | 13–18 | Good work | 19–24 | Excellent! |

For more information on this topic, see pages 104–105 of your Success Revision Guide.

Telescopes and Astronomy

Multiple-choice questions

Choose just one answer: A, B, C or D.

1. A refracting telescope has an objective with a power of 0.3 D and a eyepiece with a power of 6 D. What is the magnification achieved? (1 mark)
   - A 2
   - B 15
   - C 20
   - D 6.3

2. In their search for evidence of intelligent, extraterrestrial life in distant star systems, what are astronomers looking for? (1 mark)
   - A patterns in radio signals received from space
   - B fluctuations in light intensity from stars
   - C absorption spectra in light received on Earth
   - D replies to signals transmitted from Earth

3. Detection of very faint objects in space is improved by using: (1 mark)
   - A an objective with a greater focal length
   - B a converging lens instead of a mirror
   - C an objective of greater diameter
   - D greater magnification (1 mark)

4. What is the term for the apparent motion of stars against a more distant background?
   - A parallel
   - B parallax
   - C parsec
   - D SETI

5. Star A is 3 parsecs from Earth. Star B is 5 parsecs from Earth. Which statement best describes how these stars will appear when observed over a 6 month period? (1 mark)
   - A neither star will have moved
   - B star B will have moved more than star A
   - C star A will have moved more than star B
   - D both stars will have moved the same distance

Score / 5

Short-answer questions

1. A refracting telescope consists of two lenses arranged so that the image formed by the objective lens occurs at the focus of the eyepiece lens.

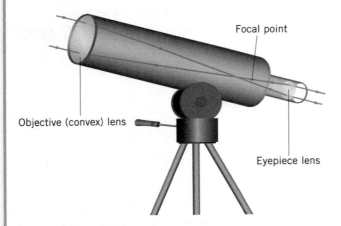

   a) Describe how you can you tell from the diagram that the eyepiece is the stronger of the two lenses. (2 marks)

   b) If the power of the objective lens is 0.75 D and that of the eyepiece is 12 D, calculate the magnification achieved. (1 mark)

   c) Calculate how long the telescope would be. (3 marks)

Score / 6

GCSE-style questions

Answer all parts of all questions. Continue on a separate sheet of paper if necessary.

1 The diagram shows a reflecting telescope.

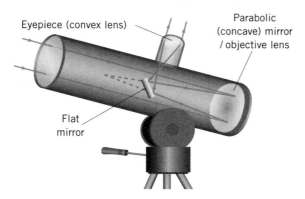

a) Describe how the light collection in a reflecting telescope differs from that in a refracting telescope. **(1 mark)**

b) Suggest two advantages of this telescope design over the use of a refracting telescope. **(2 marks)**

c) Professional telescopes tend to be very large. For example, the Keck telescope is almost 25 m in length. What advantages does an increase in size provide, and why? **(3 marks)**

2 As the earth orbits the Sun throughout the year, the positions of objects in space that are relatively close appear to change against the background of more distant objects. The same effect can account for experimental errors when taking readings in the laboratory if these are not read at the correct angle.

a) State whether the parallax angle is greater or smaller if the object is more distant. **(1 mark)**

b) Arcturus is the brightest star in the constellation Boötes. It has an angle of parallax of 0.089 seconds of arc. Describe the relationship between the angle of parallax and the parsec (the unit of measurement used by astronomers. **(1 mark)**

Score / 8

How well did you do?

| 0–4 Try again | 5–9 Getting there | 10–14 Good work | 15–19 Excellent! |

For more information on this topic, see pages 106–107 of your Success Revision Guide.

Electromagnetic Effects

Multiple-choice questions

Choose just one answer: A, B, C or D.

1 Which of the following changes would reduce the force on a current-carrying wire held in a magnetic field? **(1 mark)**
- A increasing the strength of the magnets
- B turning the wire to be more aligned with the magnetic field
- C using thinner wires with the same current
- D using a longer wire

2 When using Fleming's left hand rule what does the thumb represent? **(1 mark)**
- A the magnetic field
- B the movement of electrons along the wire
- C the force on the wire
- D the magnetic field around the wire due to the current flowing

3 In a d.c. motor, what is the split-ring commutator designed to do? **(1 mark)**
- A reverse the direction of current in the coil once every turn
- B increase the force on the coil
- C enable a moving electrical contact
- D reverse the direction of current in the coil every half turn

4 A transformer has 970 turns on its primary coil and an output voltage of 5 V a.c. when connected to the mains. How many turns are there on the secondary coil? **(1 mark)**
- A 21
- B 194
- C 50
- D 143

Score / 4

Short-answer questions

1 When a current passes through a wire held in a magnetic field, the wire experiences a force that makes it move.

a) What is the name of this effect? (1 mark)

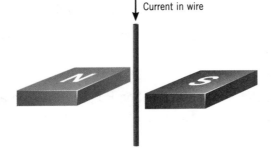

b) On the diagram above, draw an arrow to show the direction of the magnetic field in the gap between the poles. Label this arrow A. (1 mark)

c) On the diagram above, draw a second arrow to show the direction in which the wire will be forced. Label this arrow B. (1 mark)

d) What effect, if any, will reversing the direction of the current have? (1 mark)

Score / 4

GCSE-style questions

Answer all parts of all questions. Continue on a separate sheet of paper if necessary.

1 A generator consists of a coil of wire that is rotated in a magnetic field. The coil has 20 turns of wire. Moving contacts at each end of the coil are connected to the input of an oscilloscope to display the induced voltage.

  a) On the grid alongside, sketch what would be seen on the oscilloscope output if the coil was rotated four times a second. Indicate an appropriate time scale. **(3 marks)**

  b) Explain the effect(s), if any, of the following changes:

   i) Doubling the rotation rate **(2 marks)**

   ii) Turning the coil in the opposite direction **(2 marks)**

   iii) Using 40 turns of wire on the coil **(2 marks)**

2 A step-down transformer in an adaptor connects an electrical device to the mains electricity supply and provides a voltage of 12 V.

  a) Describe what is meant by the terms 'primary' and 'secondary' in connection with a transformer. **(2 marks)**

  b) State the relationship between the number of turns of wire on the primary and secondary coils. **(1 mark)**

  c) Calculate the turns ratio for a transformer at an electricity sub-station, with an input of 33 kV and an output of 230 V. **(1 mark)**

Score / 13

How well did you do?

0–5 Try again | 6–11 Getting there | 12–17 Good work | 18–21 Excellent!

For more information on this topic, see pages 110–111 of your Success Revision Guide.

Kinetic Theory

Multiple-choice questions

Choose just one answer: A, B, C or D.

1 Liquid nitrogen boils at −196 °C. What temperature is this on the Kelvin scale? **(1 mark)**
- A 96 K
- B 77 K
- C 57 K
- D −77 K

2 The temperature at which an ideal gas exerts no pressure is known as: **(1 mark)**
- A the transition temperature
- B the null pressure point
- C absolute zero
- D its freezing point

3 In a gas, the average kinetic energy of the particles is proportional to: **(1 mark)**
- A pressure
- B Kelvin temperature
- C heat energy
- D the rate of collisions

4 When a solid is heated, the temperature stops rising whilst it melts. Why is this? **(1 mark)**
- A heat energy is being used to break bonds between particles
- B bond rearrangement within particles
- C particles evaporating
- D heat energy transfer to the surroundings

5 A gas exerts a pressure of 50 kPa. What force is exerted on the walls of the container if the surface area is 0.5 m²? **(1 mark)**
- A 25 N
- B 10 N
- C 25 000 N
- D 100 N

Score / 5

Short-answer questions

1 The graph shows how the temperature of a substance changes as it cools from being a gas to a solid.

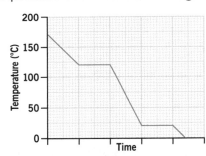

a) On the graph, label the regions where the substance is **i)** a solid **ii)** a liquid **iii)** a gas. (3 marks)

b) Use the graph to determine the melting point and boiling point of the substance. (2 marks)

   i) Melting point = ... ii) Boiling point = ...

c) Convert your answers from **b)** to the Kelvin temperature scale. (2 marks)

   i) Melting point = ... ii) Boiling point = ...

d) The latent heat of melting for ice is 334 kJ/kg. Calculate how much heat energy is required to melt 20 g of ice. (1 mark)

Score / 8

GCSE-style questions

Answer all parts of all questions. Continue on a separate sheet of paper if necessary.

1 Gases consist of individual particles in random motion. If a gas is in a fixed container then gas particles will collide with the container walls. This can be modelled by throwing a ball against a wall. If the kinetic energy of the ball is conserved during the collision with the wall, the ball rebounds at the same speed but in the opposite direction.

a) State which property of a gas is modelled by the conservation of kinetic energy when the ball strikes the wall and rebounds off it. **(1 mark)**

b) Calculate the change of momentum if a ball of mass 50 g is thrown at a wall at a speed of 5 m/s and rebounds at the same speed. **(2 marks)**

c) i) Explain why a force is exerted on the wall (and on the ball). **(2 marks)**

ii) Calculate this force if the contact time is 0.1 s. **(1 mark)**

d) Describe how the 'ball and wall' model helps to explain the pressure of a gas. **(1 mark)**

e) By considering how frequently gas particles would hit the container wall, predict the effect on pressure of reducing the volume of the container holding the same amount of gas. **(2 marks)**

2 Copper sulfate can be crystallised from its solution by leaving a sample in an evaporating dish. At room temperature (approximately 22 °C) all of the water will evaporate within a few days.

a) Explain what 'evaporation' is and how it differs from 'boiling'. **(2 marks)**

b) When our bodies overheat we lose water at the skin surface by sweating, which then evaporates. Why does evaporation have a cooling effect? **(1 mark)**

Score / 12

How well did you do?

| 0–6 Try again | 7–12 Getting there | 13–18 Good work | 19–25 Excellent! |

For more information on this topic, see pages 112–113 of your Success Revision Guide.

The Gas Laws

Multiple-choice questions

Choose just one answer: A, B, C or D.

1 Which of the following is NOT proportional to temperature in an ideal gas? **(1 mark)**
  A volume
  B mass
  C pressure
  D average particle kinetic energy

2 What is 42°C when converted to kelvin? **(1 mark)**
  A 420 k
  B 84 K
  C 315 K
  D 414 tK

3 Which of the following statements is false when explaining why gas pressure increases with temperature at constant volume? **(1 mark)**
  A there are fewer particle collisions
  B the particles have more kinetic energy
  C there are more collisions with the walls
  D there are more frequent, harder collisions with the container walls

4 A gas at constant volume is put under twice the pressure. What happens to the average kinetic energy of the gas molecules? **(1 mark)**
  A it doubles
  B it increases by 50%
  C it halves
  D there is no change

5 The pressure of an ideal gas doubles and its temperature halves. What happens to the volume? **(1 mark)**
  A it is reduced four-fold
  B it is halved
  C it stays the same
  D it is doubled

Score / 5

Short-answer questions

1 A pressurised paint spray can contain liquid paint and gas. When the nozzle is depressed, the gas forces some of the paint out. Some of the gas escapes with the paint, creating an aerosol.

  a) Describe what causes the pressure in the can. (1 mark)

  b) As the paint is used up it becomes more difficult to get a good spray from the can. Explain why this is. (2 marks)

2 A cyclist pumps up her deflated tyre by injecting more gas particles (air molecules).

  a) Describe how this increases the gas pressure in her tyre. (1 mark)

  b) Later in the day the cyclist checks the tyre pressure again and finds that it has increased. State which property of the gas could have changed to bring about this increase in pressure (1 mark)

Score / 5

GCSE-style questions

Answer all parts of all questions. Continue on a separate sheet of paper if necessary.

1 A student carries out an investigation into the effects of volume and temperature on gas pressure. He draws graphs of the results.

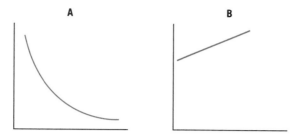

a) State which graph represents each of the following relationships. (2 marks)

   i) volume of gas and gas pressure. _____

   ii) temperature of gas and gas pressure. _____

b) State whether the student plotted the temperature graph in °C or K and explain how you can tell. (2 marks)

2 An empty 500 ml plastic drinks bottle is heated in boiling water and sealed with the air inside at 80°C. It is then removed and allowed to cool to room temperature (20°C), during which time it collapses. There is no change in pressure.

a) Calculate the volume of gas inside the bottle at the end of the experiment. (2 marks)

b) Calculate the air temperature in the bottle when the volume of gas had fallen to 450 ml. (2 marks)

3 A rigid, sealed container of gas is heated from room temperature and pressure until the pressure becomes 400 kPa. Atmospheric pressure is 100 kPa. There is no change in volume.

a) Calculate the temperature of the gas. (2 marks)

b) Heating continues until 1200°C. Calculate the pressure of the gas at this temperature. (2 marks)

Score / 12

How well did you do?

0–5 Try again | 6–11 Getting there | 12–17 Good work | 18–22 Excellent!

For more information on this topic, see pages 114–115 of your Success Revision Guide.

Medical Physics

Multiple-choice questions

Choose just one answer: A, B, C or D.

1 Which of the following is an advantage of ultrasound medical imaging over the use of X-rays? (1 mark)
- A it is more sensitive
- B it does not involve ionising radiation
- C it is less penetrating
- D it is very efficiently reflected by bone

2 Which one of the following is NOT an application for ultrasound? (1 mark)
- A destruction of kidney stones
- B investigating metal fatigue
- C computer networking
- D measuring blood flow

3 Which of the following does NOT use echolocation with sound or ultrasound waves? (1 mark)
- A bats
- B air-traffic controllers
- C submarine operators
- D traffic speed cameras

4 Which of the following radioisotopes is commonly used for PET scans? (1 mark)
- A carbon-14
- B potassium-40
- C fluorine-18
- D phosphorous-32

5 Which of the following body structures would give the strongest ultrasound echo? (1 mark)
- A the rib cage
- B the liver
- C the abdominal muscle wall
- D the heart

Score / 5

Short-answer questions

1 Fill in the missing words to complete the passage below. Select from the words provided. (7 marks)

computerised   annihilate   gamma   targeting   glucose   ultrasound   protein

tomography   radiopharmaceutical   tumours   X-rays   combine

Positron emission .... (PET) can be used to find the positions of .... by identifying areas of high .... metabolism. First, the patient is given a ...., which is a biologically active molecule labelled with a positron-emitting isotope, e.g. a version of glucose labelled with fluorine-18. This collects in areas of high glucose metabolism. Positrons emitted at this site .... with electrons and produce a pair of .... rays in opposite directions. The computer produces an image showing where in the body the radiation came from. PET scans can be combined with scans using .... (CT, or CAT, scans) in order to relate this to anatomical structures.

Score / 7

GCSE-style questions

Answer all parts of all questions. Continue on a separate sheet of paper if necessary.

1 PET is an imaging technique used in hospitals that relies on the nuclear decay of isotopes, such as fluorine-18, which releases positrons.

   a) Describe what a positron is. (2 marks)

   b) Complete the following nuclear equation for the decay of fluorine-18: (2 marks)

   $$^{18}_{9}F \rightarrow ^{\Box}_{8}O + ^{0}_{\Box}\beta^{+}$$

   c) The half-life of fluorine-18 is 110 minutes.
   i) Describe what practical issues this half-life presents in the use of fluorine-18 for imaging in hospitals. (1 mark)

   ii) In terms of the patient, describe what advantage there is to using an isotope with such a short half-life. (1 mark)

   d) PET works by detecting the oppositely-directed gamma rays produced when positron-electron annihilation takes place.
   i) Describe what is meant by 'positron-electron annihilation'. (2 marks)

   ii) Explain why annihilation results in paired gamma rays. (1 mark)

2 Ultrasound can be used in tumour diagnosis. The ultrasound is produced as a short pulse. Echos are produced if it strikes dense matter, e.g. a tumour. The echo is received after a short time delay. An image is built up by scanning a narrow beam across the tissue under examination.

   a) The speed of sound in the body is 1500 m/s. Calculate the time delay for an echo from a tumour at a depth of 5 cm. (2 marks)

   b) Some of the ultrasound transmits through the tumour and echoes off the boundary at the far side. If the tumour is 2 cm across, calculate how much later the second signal would be received. (2 marks)

Score / 13

How well did you do?

| 0–6 | Try again | 7–12 | Getting there | 13–18 | Good work | 19–25 | Excellent! |

For more information on this topic, see pages 116–117 of your Success Revision Guide.

Notes

Notes

Notes

Notes

Notes

Notes

Periodic Table

Key

relative atomic mass
atomic symbol
name
atomic (proton) number

1	2												3	4	5	6	7	0
						1 **H** hydrogen 1												4 **He** helium 2
7 **Li** lithium 3	9 **Be** beryllium 4												11 **B** boron 5	12 **C** carbon 6	14 **N** nitrogen 7	16 **O** oxygen 8	19 **F** fluorine 9	20 **Ne** neon 10
23 **Na** sodium 11	24 **Mg** magnesium 12												27 **Al** aluminium 13	28 **Si** silicon 14	31 **P** phosphorus 15	32 **S** sulfur 16	35.5 **Cl** chlorine 17	40 **Ar** argon 18
39 **K** potassium 19	40 **Ca** calcium 20	45 **Sc** scandium 21	48 **Ti** titanium 22	51 **V** vanadium 23	52 **Cr** chromium 24	55 **Mn** manganese 25	56 **Fe** iron 26	59 **Co** cobalt 27	59 **Ni** nickel 28	63.5 **Cu** copper 29	65 **Zn** zinc 30	70 **Ga** gallium 31	73 **Ge** germanium 32	75 **As** arsenic 33	79 **Se** selenium 34	80 **Br** bromine 35	84 **Kr** krypton 36
85 **Rb** rubidium 37	88 **Sr** strontium 38	89 **Y** yttrium 39	91 **Zr** zirconium 40	93 **Nb** niobium 41	96 **Mo** molybdenum 42	[98] **Tc** technetium 43	101 **Ru** ruthenium 44	103 **Rh** rhodium 45	106 **Pd** palladium 46	108 **Ag** silver 47	112 **Cd** cadmium 48	115 **In** indium 49	119 **Sn** tin 50	122 **Sb** antimony 51	128 **Te** tellurium 52	127 **I** iodine 53	131 **Xe** xenon 54
133 **Cs** caesium 55	137 **Ba** barium 56	139 **La*** lanthanum 57	178 **Hf** hafnium 72	181 **Ta** tantalum 73	184 **W** tungsten 74	186 **Re** rhenium 75	190 **Os** osmium 76	192 **Ir** iridium 77	195 **Pt** platinum 78	197 **Au** gold 79	201 **Hg** mercury 80	204 **Tl** thallium 81	207 **Pb** lead 82	209 **Bi** bismuth 83	[209] **Po** polonium 84	[210] **At** astatine 85	[222] **Rn** radon 86
[223] **Fr** francium 87	[226] **Ra** radium 88	[227] **Ac*** actinium 89	[261] **Rf** rutherfordium 104	[262] **Db** dubnium 105	[266] **Sg** seaborgium 106	[264] **Bh** bohrium 107	[277] **Hs** hassium 108	[268] **Mt** meitnerium 109	[271] **Ds** darmstadtium 110	[272] **Rg** roentgenium 111							

Elements with atomic numbers 112–116 have been reported but not fully authenticated

*The lanthanoids (atomic numbers 58–71) and the actinoids (atomic numbers 90–103) have been omitted.
The relative atomic masses of copper and chlorine have not been rounded to the nearest whole number.